Agricultural Modernization and Social Change in Indian Villages

OTHER BOOKS

- आदिवासी विकास: उपलब्धियां एवं चुनौतियाँ (भाग-1) एस. एन. चौधरी और मनीष मिश्रा (सम्पादक)
- आदिवासी ग्रामों में टिकाऊ विकास का यथार्थ/एस.एन. चौधरी, मनीष मिश्रा एवं मनोज शर्मा (सम्पादक)
- Dalit Agenda and Grazing Land to Scheduled Castes and Scheduled Tribes/ S.N. Chaudhary
- Dalit and Tribal Leadership in Panchayats/ S.N. Chaudhary
- Human Rights and Poverty in India: Theoritical Issues and Empirical Evidences (in 5 Volumes)/ S.N. Chaudhary (Ed.)
- Tribal Development Since Independence/S.N. Chaudhary (Ed.)
- Tribal Education: Implications for Development/ S.N. Chaudhary (Ed.)

About the Author

S.N. Chaudhary, Professor of Sociology and former Dean, Faculty of Social Sciences, is currently Rajiv Gandhi Chair Professor at Barkatullah University, Bhopal, Madhya Pradesh. He has also been a fellow of the A.N. Sinha Institute of Social Studies, Patna, Indian Institute of Advanced Study, Shimla and SMH Paris. He has also conducted fieldwork among Red Indians in British Columbia (Canada). Dr. Chaudhary produced 31 Ph.Ds., completed 28 research projects and published 32 books and 60 research articles. Currently, Professor Chaudhary is working on Trafficking of Tribal Girls and Women, Impact of Technology on Society, Suicide among Tribal Farmers and Agricultural Labourers and Tribal Development.

Agricultural Modernization and Social Change in Indian Villages

S.N. Chaudhary

CONCEPT PUBLISHING COMPANY PVT. LTD.
NEW DELHI-110059

ISBN-13 : 978-93-5125-044-9

First Published 2014

Published and Printed by

Concept Since 1974

Concept Publishing Company Pvt. Ltd.
A/15-16, Commercial Block, Mohan Garden
New Delhi-110059 (India)
T: 25351460, 25351794, *F*: 091-11-25357109
E: publishing@conceptpub.com, W: www.conceptpub.com
Editorial Office: H-13, Bali Nagar, New Delhi-110 015, India.

Cataloging in Publication Data--*Courtesy:* D.K. Agencies (P) Ltd. <docinfo@dkagencies.com>

Chaudhary, S. N. (Shyam Nandan), 1954-
Agricultural modernization and social change in Indian villages / S.N. Chaudhary.
p. cm.
Includes bibliographical references (p.) and index.
ISBN 9789351250449

1. Agriculture--Economic aspects--India. 2. Rural development--India. 3. Social change--India. 4. Krishi Vigyan Kendra. I. Title.

DDC 338.10954 23

Preface

History of debate on technological modernization in agriculture and its impact on the economic life of farmers in particular is old in India. However, with political independence in 1947 and introduction of new democratic constitution and number of area and target groups specific development projects in the subsequent years, discussion on rural development through agricultural modernization became more dominant, intense and deep-rooted. Improvement approach initiated during the British period for the so-called rural development became dominant approach during the free India. It was thought by the dominant economic and political order that only through the adoption of modernization model of development in agricultural management nation can meet ever increasing food requirement of the increasing population. Community Development Project was one of the institutional responses to this question. But this and similar projects had limitations. Nation as such had to depend on import of food grains. This led to the birth of Green Revolution. Although green revolution and its production based impact was not uniform in different parts of the country but at the national level it was a grand success. The problem of food insecurity was now the thing of the past.

But this does not mean that success of green revolution in one area led to success in other realms. Economic disparity and subsequently violence increased among regions, social groups and communities. It was realized through different impact studies that technological modernization in agriculture through C.D. Projects, Green Revolution and such other interventions definitely brought huge farm production which inturn contributed to economic development and subsequently development in other walks of life for small number of people who were already medium or big land-owners but its benefit was almost notional or insignificant for the poor and the marginalized. In many cases poor became poorer and poorer.

From social change point of view impact of modern farm technology was also not uniform. In some areas it contributed to just and democratic values but in many other areas it reinforced tradition. Economic development and dowry went hand in hand. Also, impact was not free from gender-bias. In many parts of the country Green Revolution led economic development contributed to caste and class conflict and rural violence.

So far as Krishi Vigyan Kendra (KVK) is concerned it is one of the four interventions to make sure transfer and proper use of modern farm technology for high farm production and economic development at the village level. It is also most viable intervention for change in the mind-set and capacity of farmers. How and to what extent it has succeeded in its agenda? How economic development have adjusted with democratic dimensions of social change? What is the future of this process? How it has impacted the cropping pattern? To what extent it is poor or farm wage earner or landless friendly? These are some of the broad issues which have been covered in this study.

In order to undertake and conduct this study we received support and cooperation from different organizations and individuals. It is our moral and social responsibility to acknowledge all of them.

At the very outset I am highly obliged and thankful to ICSSR, New Delhi in general and its concerned officials in particular for sanctioning the project and releasing the grant in time. Without their support it would have been rather impossible to conduct the study in the remote rural villages of M.P.

Shri Manoj K. Sharma needs appreciation for his committed performance in data collection work as a Research Fellow. I also, extend thanks to Shri Arvind Saxena, Research Scholar who helped not only in data collection from the field but also he brought huge relevant and related materials from secondary sources.

Last but not the least, I am thankful to the villagers and KVK officials who spared time for interview and provided required data and information for successful completion of the study.

July 2013 **S.N. Chaudhary**

Contents

List of Tables

List of Figures

CHAPTER – ONE

Introduction

Background

Historical records clearly reflects that no underdeveloped countries have moved from the stage of chronic poverty and stagnation to the stage of economic development and progress without achieving a substantial increase in agriculture production. In the earlier stages of economic development of modern advanced states, a high rate of agriculture production has played a crucial role in furthering overall economic growth (Nichols, 1964). So far as India is concerned it is a land of villages. Mahatma Gandhi rightly said that India lives in villages. And even today about 70 per cent population lives in villages. Their main source of living is agriculture and allied activities. In other words, the rural character of India's economy is highlighted in the proportion of people living in rural areas and their dependence on economic activities based on primary sector for their livelihood. This is also evident from the fact that,

> "Indian agriculture employs 61.0 per cent of economically active population, which is higher by 15 per cent than the world average 46.0 per cent. India shares 2.4 per cent of world total area (seventh rank) and 2.3 per cent of world land area, 1.8 per cent of world arable land and 23.6 per cent of world irrigation area with these resource, the country is to support 16.4 per cent of world population and 21.1

per cent of world population who depend on agriculture". (FAO,1997).

India has the second largest arable land in the world next only to U.S.A. The geographical area of India is equivalent to the total geographical area of 15 countries of the European Economic Country (EEC). India can feel proud to have encompassed within its geographical wealth of the 15 EE countries. During the winter months, the entire Europe reels under tight grip of snow and no agriculture crop can be cultivated. But a large number of agriculture and horticultural crops can be cultivated throughout the year in India which can be profitably exported to other markets. India is in proud possession of third largest reservoir of scientific and technical manpower in the world. (Mahalingam, 2002).

Agriculture being the mainstay of the rural Indian economy around which socio-economic privileges and deprivations revolve, any change in its structure is likely to have a corresponding impact on the existing pattern of social inequality. According to Kuppuswamy (1990), agriculture is the mainstay of the rural economy and so the entire economy and social structure revolve around it. All the rural occupations are generally integrated with agriculture on the basis of the caste system and a caste based *jajmani* system. Farmers belonging to the carpenter caste or blacksmith caste or the potter caste, those belonging to the service caste like the barbers and the washerman and the landless labourers belonging to the Scheduled Castes, all have obligations to serve the needs of the farmers. For the service rendered, the farmer gives a share of the crop as fixed by tradition. Thus, most of the economic transactions in the villages have been historically based on exchange of goods or services or both. Sociologically this practice is known as *jajmani* system. This system refers to organic relationship between and among different castes and sub-castes. It is needless to mention that historically land has been the main and sometimes the only source of livelihood in the villages. However, most parts of the land have been owned by numerically inferior but otherwise dominant section of the society, most of whom come from upper castes. They owned land and determined the patterns of its use. The "others" particularly the intermediate and the Scheduled Castes were dependent on land as a sharecropper. Many of them were marginal farmers. And most of the wage earners were from the Scheduled Castes, particularly in the non-tribal areas.

Hence, in the villages people across their background were directly or indirectly dependent on land for their livelihood, and social status.

Rural development in India is equated with the development of the vast majority of farmers. The major aim of rural development efforts is to bring about improvement in the physical and social quality of life of rural people. Life in rural areas, by and large, revolve around agriculture. Unless agriculture is developed, other spheres of rural life cannot prosper. Growth and development of agriculture and rural life, therefore, are inseparable. Nair (1999) rightly said that agriculture is the mainstay of more than 65 per cent of India's population. The agriculture sector accounts for about 28 per cent of the Gross Domestic Product (GDP). Farmers contribute substantially to the agriculture production, particularly in the matter of foodgrains. The phenomenal increase in the production of foodgrains from 51 million tonnes in 1950-51 to 199 million tonnes during 1996-97 helped the country to become self-sufficient in foodgrains. In other sectors of the agriculture economy too there has been a substantial increase in the production of various commodities such as oil, fruits and vegetables, eggs, fish and milk.

Rural society is generally comprehended in contrast with the urban society, the two concepts having assigned positions at the opposite poles of a continuum. Some analysts have tried to conceive rural and urban society in terms of a dichotomy, this assigning apparent exclusiveness to their character. Broadly speaking, a rural area can be delineated in-terms of a geographical space, which is outside the limits of any industrial or urban setting. This kind of delineation only in-terms of space as such remains unsatisfactory until other important attributes are taken into account. Whether in-terms of urbanity of geographical space, a rural area is a natural region having a set of characteristic network and other flora, fauna and cultural attributes. None of the characters as such may incorporate several of them in various combinations.

In the Indian context, by rural area we understand to be of non-municipal areas of the country side which, by and large, are represented by the agricultural productive system and which lack certain urban amenities, proportion of literacy being low and having insignificant industrial growth, various human population in rural India are relatively homogenous and live in relatively closer harmony with the nature.

Finally, it can be said that agriculture has left a significant impact on the socio-economic activities of Indian society. Farmers have manifested themselves through the increased yield levels, significant changes in farming and non-farming items, modified credit behaviour, market behaviour and storage patterns.

Concept of Agricultural Modernization

Modernization at the individual level corresponds to development at the societal level. It is a process by which individuals change from a traditional way of life to more complex, technologically advanced, and rapidly changing style of life. According to Singh (1976), there are four bipolar formulations of the concept of modernization. These are psychological, normative, structural and technological formulation. All these formulations are related to each other. All have contextual meanings which includes elements of history, culture, social structure, and existential factors. So far as technological formulation of modernization is concerned it is analyzed in terms of economic resources, or use of inanimate power (here replacement of ox and plough by tractor or use of chemical fertilizer instead of bio-fertilizer). To him, in such formulations modernization is associated with material inputs, and developmental infrastructures, which bring about qualitative and progressive mobilization in the total resources of society (p. 55).

In mid-1960s agriculture in India was in a condition of great difficulty. During 1950s, and the early 1960s agriculture growth in India was sustained mainly by expansion of area under cultivation. But by mid-1960s the scope for increasing the area under cultivation has become fairly limited, the growth of yield rate of crops was already very weak and there was not much scope to increase agriculture or farm work without some major technological break through in the farm sector. Hence, new inputs in the form of new knowledge, new technology, fertilizer, water management techniques, use of HYV seeds, etc., became essential. Keeping these objectives in mind new technology was introduced in a phased manner. In other words, the new technology which gradually replaced the old mode of cultivation was based on some newly developed crop varieties known as the High Yielding Varieties (HYV) of seeds. The new agricultural strategy aimed at increasing agriculture production in the country by replacing the old method of farming by a modern and more efficient technique of cultivation.

Indian agriculture has begun to reap forward from the traditional mode to that of sustainable agriculture by using new technology for production. The new strategy did succeed in boosting productivity and sustaining production growth in agriculture of the country in the years that followed. As their name suggests those of traditional local varieties of seed, provided of course, their use was supported adequately by complementary inputs such as chemical fertilizer and pesticides and controlled watering. However, as the adoption of HYV is not the only indicator of agricultural modernization, it would not have vitiated the process significantly. One can take the question of land reform, consolidation of farm land to the landless and ceiling on upper limit of land. Though the concept of agricultural modernization is broad and complex, it refers to the extensive and intensive use of improved production technology and inputs for the maximization of production.

Traditional farming has supported human population on the same lands for thousands of years. The modern agricultural practices include monoculture of crops, heavy and repeated application of chemical fertilizers, pesticides to combat wide array of insects and pathogens and substitution of machinery for human and animal labour. It also requires ensured and sufficient irrigation facility. Its features are broadly in direct contrast to our traditional farming systems which have evolved, expanded and survived in the absence of all these inputs for centuries. In modern agriculture, with the sole objectives of maximizing field, number of other desirable characteristics are required.

The discussion relating to traditional ideas, religious beliefs and traditional social institutions, such as the caste system, raise the important question: can a superstructure of modernization be built on a foundation which still retains some of the essentials of the old social and economic order? It is undoubtedly true that all traditional systems can absorb traits of modernization without visible injury but we would like to emphasize the point that the stage soon be reached where a traditional social order and the social requisites of far-reaching economic development and technological change become irreconcilable. Let us take the case of the institutional frame of Indian agriculture to provide an illustration of the point. Though our agrarian structure has proved capable of absorbing a limited measure of change as illustrated by the limited success of the new agricultural strategy introduced since late 1960s, modernization of the bulk of the farm population still hinges on how quickly that frame is thoroughly overhauled. For, it is the perpetuation of this agrarian structure that has acted as the greatest stumbling block to agrarian development and has, in effect, perpetuated the retrograde traditional ideas and institutions. It seems that India faces, at this juncture of development, the following critical choice:

(a) Should we deliberately regulate, with a view to preserving certain specified elements of the traditional order, the content, direction and speed of modernization? or
(b) Should we deliberately regulate the traditional order so as to attain certain specified goods of Modernization? (Dube, 1974)?

While making the choice we should do well to remember that no traditional society has been found to survive the process of technological modernization and economic development without making some dent into its traditional order. As such, developing countries like India are faced with a real dilemma: if they want to preserve some elements of the traditional order, they can do so only at the cost of possible retardation of economic growth. And if they want to achieve a measure of economic and technological modernization, they cannot but allow injury to be caused to the traditional order.

And that traditional order is informant. By accentuating the difference between the haves and have-nots, by pushing an ever increasing number of people below the level of subsistence, that traditional social order has given rise to social tension of a magnitude capable of generating far reaching change in the socio-economic and political structure of India. Should we go in for orderly change or should we allow things to their own course? This is the question before the Indian leadership today. The pertinent challenge before all of us is that we want economic development but we do not like to compromise on social and environmental issues. We want food to everybody at our own and not through import but we do not want our natural resources to die. We want technological modernization in the field of agriculture but we also want work to every hand. How it is possible? Most of the grassroots based empirical studies support the thesis that development has both functional and dysfunctional implications. It is full of gains and pains.

The study of history of agriculture development in India and the role of planning in that development convinces us that certain events and situations have greatly influenced development since independence. One of such events and situations is the availability of land for an expansion in the gross areas cropped in 1950s and the non-availability of such land in the 1960s. The condition or event sustained agriculture development is the 1950s and caused it to languish in the early 1960s. 'From a very significant growth of 1.8 per cent a year during the first ten years after independence, the growth rate in output attributable to area expansion had come down to an insignificant 0.12 per cent (Minhas, 1969).

On the other hand, the rate of growth of double cropped area did not pickup appreciably in the sixties. In view of the fact that agriculture development virtually remained stagnant in British India during 1900-1947 except in occasional year of favourable weather and the area under cultivation did not increase appreciably after 1910, the extent of achievement in the first fifteen years of planning might not have taken place without, of course, the backing and organization currently being provided by a state sponsored economic plan. Bhattacharya (1970) opines that the evidence, therefore, decidedly establishes that the planned economy with its emphasis on agricultural extension, credit and improved agriculture, inputs has brought about a remarkable expansion in all these directions. But as the expansion in agriculture output so far has been much faster than the growth in population, the balance between population and agriculture resources remained almost as precarious even after 15 years of planning as before that period. The physical and financial limitations of settlement programme turned the attention of many governments in the developing countries to the possibilities of technological progress as a means to create more jobs as well as to provide surplus from agriculture. Technical progress can, by intensifying agriculture, help grow higher yields and more crops per acre and may result in more employment. Now agriculture can be intensified in two ways:

(a) by adopting bio-chemical technical advances, and
(b) by having resort to mechanization to speed up certain tasks in the peak season by way of removing blockages.

Despite the importance of the new high-yielding varieties of wheat in intensifying agriculture and augmenting the use of land, little data have been collected on its employment effects. The Food and Agriculture Organization (FAO) has estimated that a doubling in yields results in 37 per cent additional requirements of labour with the range varying between 25 and 60 per cent (Hrabovsky, 1970). Hence, technological modernization is aimed to number of issues. These issues are massive increase in farm production for both consumption and market, maintenance of ecological balance and sustainability of all these issues.

Economic Development

Economic development refers to an improvement in GDP, GNP and an achievement of higher levels of real per capita income, and improvement in the living condition of people. The term also has a technical meaning according to which economic development refers to a process of economic growth within a economy, the central objectives of the process being higher and rising real per capita income for that economy.

Economic development is determined by number of essential tests; first, income must rise cumulatively for some considerable period. Second, the increase in income must bestow income benefits upon the population as a whole, covering various occupational and income groups. It is essential to recall that per capita income can rise either because added income accrues to a few persons within a highly inequitable income structure or because the same added accrues to many people within a more equitable income structure. In the domain of economic development, a rise in real per capita income comes about as a result of more than just a few receiving higher incomes. Third, as a modification of the general test that real per capita income must rise cumulatively over time, instance may exist in which economic development can still be regarded as occurring even though real per capita income does not rise.

The main goal of economic development is to raise the level of economic activity and thereby to ensure a higher standard of living both in qualitative and quantitative sense. It involves a more efficient utilization of resources and an increase in productivity by increasing the output of food and labour force required for industrial development, by expanding the market for industrial goods, by increasing its output for higher exports goods enabling the country to import so essential capital goods and by transferring a part of its surplus towards the development of industrial sector.

Any comprehensive programme of economic development will not only involve a reaching change in the economic structure but also it affects the social and psychological issues. Therefore, it is not proper to define economic development precisely or in a absolute terms. As a process which is nowhere complete and nowhere absent; it is the product of simultaneous development in many fields. While defining the term economic development it is stated that it involves in particular an increase in productivity, a more efficient use of resources to produce more than better food, clothing, shelter and other necessities and amenities of life at a less heavy cost in human toil and hardship. Economic development has far reaching implications for all aspects of life. It makes possible higher standards of nutrition, health and education. It permits an increase of leisure and opens up new intellectual frontiers. Given a wide and equitable distribution of its benefits, it is likely to result in a substantial increase in the security of the individual and in social stability.

The contribution of agriculture to economic development can be judged on the basis of what constitutes as a sector in the dualistic structure of the economy and how it limits the process of growth in other sectors of the economy. The role of agriculture in development can similarly be discussed on the basis of how it helps the development of an underdeveloped economy from the stagnant to the progressive stage.

Agriculture is by far the largest sector of the economy in most of the underdeveloped countries including India. Not only it contributes to employment opportunities to the extent of about 60 to 80 per cent of the population, but also it contributes about 50 per cent of the national income. Thus, development in agriculture is of vital importance in any scheme of economic development. The proper development of agriculture will have tremendous impact on the development of industrial sector and the overall economic development of the country.

Agriculture and Economic Development

So long as farmers continue to feed and clothe the world, agriculture continues to assume the role of a basic industry in some degree or the other in every country. In recent years, the achievements of science and technology have been tremendous in all productive fields manufacturing a variety of goods, sometimes replacing the traditional goods. But science has not succeeded in thoroughly revolutioning agriculture and replacing crops, livestock, etc. It may be rightly observed that even those most helpful in the progress of science would scarcely contend that the growing of crops and livestock will be supplanted by huge, smoky manufacturing plants producing necessities of life now furnished by farmers. Agriculture, as a primary industry, play significant role in the process of economic development of the country. But the contribution of agriculture sector to the overall economy varies from country to country depending upon the level of economic development. In the early stages of economic development, normally agriculture was the major contributor to national income. It provided employment to a majority of people. At later stages of a fairly high level of economic process, the importance of agriculture gradually declined and importance of secondary sector activities increased. However, agriculture with its functional importance has its own contribution to make throughout the unending process of economic development.

Economic development may be defined as a transformation of an economy, which is predominantly agricultural and traditional into one which is largely industrial and modern. The agricultural and non-agricultural sectors remain separate, presenting economic dualism, which in the latter stage get integrated together. In the overall economic development, the role and functions of agricultural sector have figured pre-eminently in the theories of development, particularly since Second World War.

Most of the modern developed economies have achieved development prior to First World War. In the nineteenth century and the earlier epochs, the process of economic development was spontaneous and economic theories did not devote much attention to the question of priorities, viz. whether agricultural development should precede industrial development or *vice versa*. But during the past few decades, when the Less Developed Countries (LDCs) of Asia, South America and Africa entered the phase of development, there was a strong bias in favour of rapid industrialization. Visvesvarayya (1920) a well known statesman and engineer said: "Indians will have to choose whether they will be educated or ignorant, whether they will come into closer contact with the outer world and become responsive to its influences; whether they will be organized or disunited, bold or timid, enterprising or passive, an industrial or an agricultural nation, rich or poor; strong and respected or week and dominated by forward nations" (pp. 273-74). "Industrialize or perish" model of development became a catch world after independence. Government of India accepted that through rapid industrialization and the creation of strong nation-state, India could catch up with the West. Second Five Year Plan document felt that underdevelopment was "a consequence of insufficient technological progress (GOI, 1956:6). To it this obstacle may be overcome by adopting new and heavy technology in different fields including agriculture and promoting industrialization as it has been done in the West. This was largely due to a feeling among them that industrialization was synonymous with economic development. Concentration upon accelerated industrialization became almost the motto of the less developed countries. However, the strategy that was adopted towards this end faced serious obstacles and several countries has failed in their endeavors ànd has ended in frustration.

In Western Europe, Nichols (1964) said, revolutionary changes took place during 1780–1850. Among other factors, agriculture on the eve of the Industrial Revolution, was favourable for a take-off. In the United Kingdom agriculture did provide cheap and sufficient food in the early decades of the Industrial Revolution and did provide self-finance for more efficient agriculture; but it did not, as agriculture did in Japan in later years, helps capital formation and industrial development in general. The take-off period of France, Belgium, Germany and Sweden also rested upon a firm base of a rising agriculture productivity. History of Economic Development of Russia, China and Japan reveals the basic role-played by agriculture in the earlier stages, Russian economic progress was made possible by an initially substantial food surplus. Japan with an agricultural economy, characterized by land shortage and labour surplus adopted policies which created a food surplus, the basic foundation for its spectacular economic development. Communist China also recognized the importance of achieving an agricultural surplus in the process of economic development. In India, too, during the planning era, some of the State Governments have accorded high priority to agriculture.

Myrdal (1968) also observed, "It is in the agricultural sector that the battle for long-term economic development in South Asia will be won or lost." With reference to India, Coale and Hoover (1958) stressed the same point: Very substantial progress in the most backward part of the economy (agriculture) was a prerequisite to successful development of Indian economy as a whole.

Challenges before Agriculture

Agricultural development is important for many reasons. Agriculture must primarily provide adequate food and nutrition supplies to the fast growing population. In the early stages of economic growth, the income elasticity of demand for food in underdeveloped countries is estimated to be 0.6 or higher which is twice or thrice as much as that in Europe, the United States and Japan (Mellor,1969). In developing countries like India, with low nutritional standards, any rise in income is most likely to increase the demand for food and more diversified foodstuffs. Moreover, less developed countries are experiencing population explosion. In order to supply adequate food and nutrition to the growing population, the rate of agricultural production should be higher than the rate of growth of population, which in these countries varies from 2.0 to 3.0 per cent per annum. If domestic production fails to increase adequately, the gap has to be filled up by imports, involving a heavy drain on foreign exchange resources. If the food supplies fails to expand in pace with growth of demand, the result is likely to be substantial rise in food prices and consequent inflationary situation both in rural and urban sectors.

Concept of Social Change

Change is the law of nature, what is today shall be different from what it would be tomorrow. Sociologically by social change we mean change in social structure, its elements and its functions. Actually social change is a term used to describe variations in, or modifications of any aspect of social processes, social patterns, social interaction or social organizations and social institutions.

It refers to the modification, which takes, place in the life patterns of people. It does not refer to all the changes going on in the society. The changes in art, language, technology, philosophy etc. may not be included under the umbrella of "Social Change" although all of them may lead to social change.

Social change does not refer to the change in the life of an individual or the life patterns of several individuals. It is change which occurs in the life of the entire community. Also, the process of social change is not uniform in every society or within the same society at different points of time.

Generally, it is perceived that a particular factor like technology, economic development or climatic conditions causes social change, but the reality is that social change is conditioned by multiple demographic, environmental, social and economic factors and conditions.

National revolution, establishment of a village development council, creation of a government ministry, invention of a new industrial process—all these developments may be perceived as agencies of social change. Alterations in both the structure and the function of a social system occur as a result of such processes. The structure of a social system is provided by various individual and group statuses of which it is composed. The functioning element within this structure of statuses is a role, or the actual behaviour of the individual in a given status. Status and role affect other reciprocally. For example, the status of a Minister of Education determines how the individual holding that position will act, particularly in his official capacity. To the extent that he does not operate in the prescribed manner, his status may be changed. Likewise, social function and social structure are closely linked and affect each other. In the process of social change as one is altered, so is the other.

On the basis of above analysis it may be said that modernization in agricultural operation as a process leads to economic development or an improvement in income at the individual level but from sociological point of view economic development is not an end. The end is how and to what extent, functionally or dysfunctionally, economic development leads to social change. Purpose of this study, therefore, is finally to know the process of change in social structure and social institutions.

Modernization in Agriculture before Green Revolution

As agriculture was the most important occupation, the prosperity of the country depended on prosperous agriculture and contented peasantry. Even during the period of 1921-28 when prices were rising and agriculture was in a comparatively good stage, 55 per cent of the population was living on insufficient food, getting only about

73 per cent of the minimum requirements of body for maintaining efficiency. There were 50 million people in India who never had one full meal in the whole year (Dubey, 1925). The main reason for such a state of affairs was the increasing pressure of population between 1921-31 and this increase in population further exerted more pressure on agriculture.

It was, therefore, felt that a Royal Commission would be able to bring out on surface the true nature of the circumstances, which are inhibiting the revalorizing process in rural India and would, therefore, pave the way for necessary social and legislative reforms.

The Government of India, at last, acceded to the public demand and appointed a Royal Commission on Agriculture in 1926 to make recommendations for the improvement of agriculture and the promotion of the welfare and prosperity of the rural population.

Government of India convened a conference at Shimla in October 1928 to consider and implement various recommendations of the Commission. Government found it difficult to implement immediately all the recommendations of the Commission on account of their financial implications. The report was, however, accepted as a basis on which rural reconstruction of agriculture was to proceed. The provincial governments agreed to the progressive implementation of the recommendations as and when their circumstances permitted. The Commission's proposal for reorganizing services engaged in agriculture and veterinary fields was not, however, found acceptable. The only immediate outcome of recommendations was the setting up of Impirial Council of Agricultural Research (ICAR) in 1929 in a somewhat modified form. Indian Council of Agricultural Research (ICAR) was granted very limited financial assistance and therefore, could not commence operations on the desired scale. As in the past, Government was not prepared to devote sufficient finance for improving agriculture in the country.

The growth of population, unlike advanced industrial countries, was not a consequence of rising income and living standards but was mainly the result of the reduction of death rates on account of improved public health measures against epidemics, the development of maternity and child welfare services and improvement in the treatment of hospitals, patients, etc. (Report of Famine Commission, 1945, chapter I, Part II, p. 90).

Against the rising trend of population the area under cultivation did not increase in the same proportion and showed an increase of a little more than 10 per cent only. The quinquennial average for the year ending 1946-47 was 338 million acres against 305 million acres for the year ending 1914-15. Thus, the availability of land per capita continued to decline, which is clear from following Fig. 1.1.

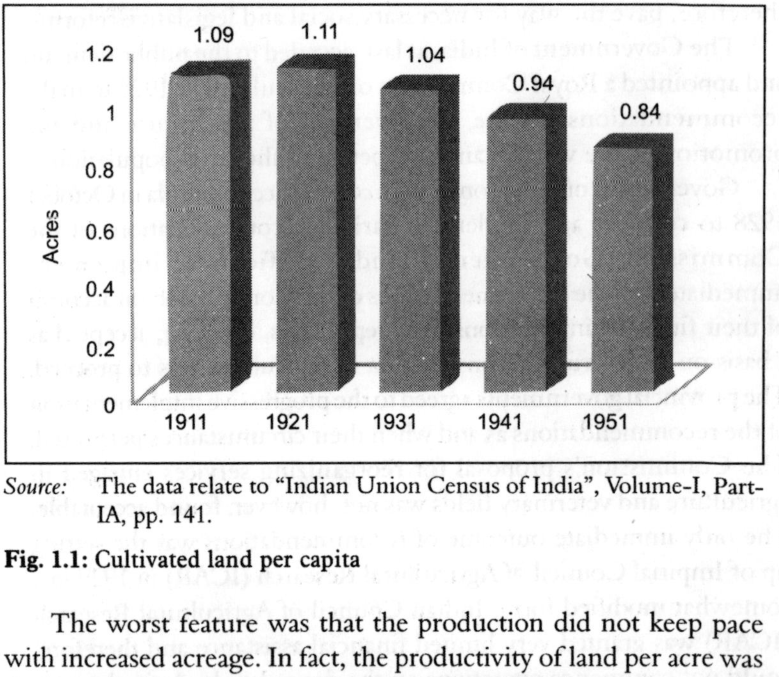

Source: The data relate to "Indian Union Census of India", Volume-I, Part-IA, pp. 141.

Fig. 1.1: Cultivated land per capita

The worst feature was that the production did not keep pace with increased acreage. In fact, the productivity of land per acre was declining. The gross value of output at 1938-39 level of prices for the quinquennium ending 1914-15 at Rs. 8109 million showed an insignificant increase and reached Rs. 8,644 million in 1946-47. The index of output at 112.5 (base year 1938-39 = 100) for the year 1910-11 was higher by about 6.36 than the figure of 106.2 for the year 1946-47. Despite high pressure of population land resources were not fully utilized. This is clear from the following observation.

Table 1.1: Area available for cultivation in 1996-99

1.	Uncultivated area	112 million acres
2.	Fallow	59 million acres
3.	Sown once a year	224 million acres
4.	Sown twice a year	36 million acres

Source: The Economic Consequences of Divided India, p.153.

Only about one-sixth of the area was double cropped while a much greater area could be sown with 2 crops in a year provided water and fertilizers could be made available. Similarly, a considerable area had to be left fallow for recuperating the fertility. There was no need to leave the land fallow if suitable rotations of crops could be introduced and adequate fertilizer provided.

The above data clearly brings out the picture of deterioration in agricultural incomes and decline in the standard of living of the cultivators and those others who were directly or indirectly dependent on agriculture.

As for the Government action for improving productivity of agriculture, it can be said without any fear of contradiction that no enough was done to put the nation's agriculture on a sound footing. Policy of agricultural betterment initiated during the regime of Lord Curzon received less and less attention as the years passed. Government attention, in the main was devoted to providing canal irrigation facilities (in Sind, Punjab, Western U.P. and Madras), encouraging the growth of credit cooperatives and security of tenures in Zamindari Provinces (Maheshwari, 1971). Provision for irrigation was a step in the right direction and it did stimulate agriculture in Sind, Punjab, Western U.P. and Madras in canal areas, but nothing was done to encourage other forms of irrigation.

The role of agriculture in income development of a country where it is most important source of livelihood, needs no description. Under the context of Indian conditions, the key role of agriculture could hardly be over-emphasized in the planned development of the economy. The country had executed Three Five Year Plans (1951-56, 1956-61 and 1961-66). During this plan period there has been a perceptible trend towards more production and better productivity in the agricultural sector. The progress over the years has, however, not been uniform as the influence of weather conditions is writ large on Indian agriculture. But each successive peak has been highest than the previous peak and each successive trough has been less than the previous trough (Mehta,1964).

There was significant increase in agricultural production during the first two Five Year Plan periods. Production index, which stood at 145.5 at the commencement of the Third Plan came down to 144.3. In a period of 16 years beginning from 1951-52 there has been about 54 per cent increase in agriculture production.

Quantitatively, the production of foodgrains, which was about 55 million tonnes at the beginning of the First Plan, reached to 82 million tonnes by the end of Second Plan. During the Third Plan the highest production of 89.36 million was recorded in 1964-65 and declined in the subsequent two years because of the unfavourable weather conditions. In 1967-68 the production reached an all time peak of 95 million tonnes and there are reports that during 1970-71 the country crossed 105 million tonnes production mark. Among foodgrains group, substantial increase in production has been recorded in wheat, maize and rice while the production of millets, gram and pulses had not recorded a satisfactory growth rate. After 1962-63 specially its production has become static or has even declined (e.g. gram, *bajra*, pulses). Production of rice and wheat at the beginning of the plan period was 22 and 7 million tonnes which reached to 39 and 12 million tonnes respectively by 1964-65. Thereafter, in subsequent 4 years rice production has become static while wheat production has increased by more than 50 per cent. Production of coarse foodgrains (*Jowar*, *Bajra* and other cereals) and pulses was 17 million and 9.18 million tonnes respectively. There was satisfactory increase in First Plan and its production jumped to 24 and 13 million tonnes by 1960-61. Since then, there has been only marginal increase in the case of coarse foodgrains while the production of pulses has actually declined from 12.71 million tonnes in 1960-61 to 10.42 million tonnes in 1968-69. Decline in gram production is specially noteworthy as its production after recording a height of 6.25 million tonnes in 1960-61 has gradually declined and stood at 4.31 million tonnes in 1968-69. This decline appears to be due to the fact that wheat cultivation has become more profitable because of its high yield due to the introduction of improved seeds and favourable prices during the past few years. The extension of irrigation facilities in Punjab, Haryana, Rajasthan and U.P. has made it possible to cultivate wheat in areas where gram was previously grown (Maheshwari, 1971).

Increase in area under cultivation has not, however, been uniform for all the crops. There has been greater increase in area under non-foodgrains as compared to foodgrains. The rate of growth in area between 1949-50 and 1968-69, under non-foodgrains has been 2.19 per cent per annum as against 1.26 per cent for foodgrains. Among the foodgrains group larger increase has been recorded for wheat and maize while sugarcane, jute, oil seeds and cotton have also been benefited by the increase in area. The area under groundnut has expanded considerably and the annual growth rate has been 3.45 per cent. The area under barley and castors has on the other hand declined substantially. Because of the success of high yielding varieties of strains, the area under wheat has increased very appreciably. The following graph shows the area under some important crops.

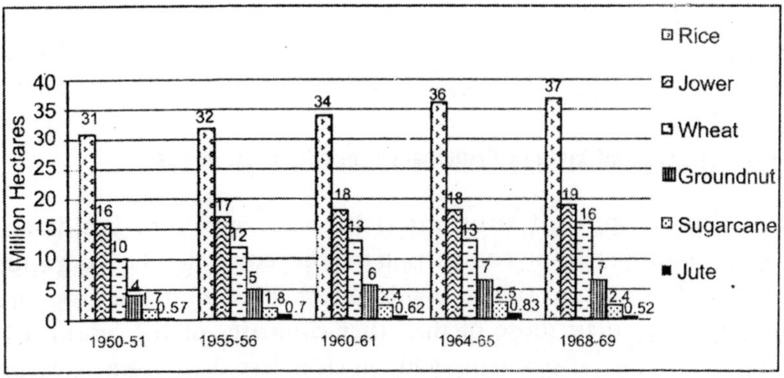

Source: Industrial and Agricultural Development of India since 1914 by B.P. Maheshwari, *Eastern Economist*, Annual Number, 1969.

Fig. 1.2: Area under main crops (million hectares)

During the first two-plan period 16.3 per cent increase in production (out of 36.4% total increase) was accounted for by higher yield per acre. Productivity has further increased specially because of the growing popularity of high yielding varieties of seeds. During 1949-50 to 1968-69 aggregate agricultural productivity rose by 1.53 per cent per annum, foodgrains at 1.65 per cent and non-foodgrains crops at 0.97 per cent.

Crop-wise largest increase has taken place in case of wheat, cotton, rice, *jowar* and tobacco. The following diagram gives the average yield of some principal crops.

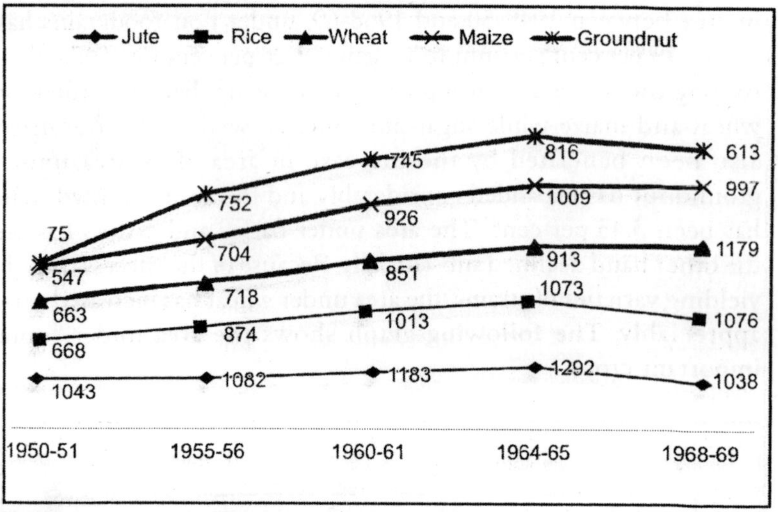

Source: *Eastern Economist* - Annual Number; 1969, p. 1337

Fig. 1.3: Average Yield of Principal Crops (Per hect. in kgs.)

But in spite of substantial increase per acre yields are disappointingly low in comparison to other countries. Leaving aside Africa, Pakistan and Burma yields in respect of all major crops are much lower than those of the other countries of the world, has almost the lowest per acre yield which is less than 1/4th of Spain and Australia and less than 1/3rd of Italy and Japan. The area under cotton is larger than in any other cotton growing country in the world. The yield per acre, however, is low at less than 100 lbs. per acre against 604 lbs. in Soviet Union, 512 lbs. in Egypt 446 lbs. in US 267 lbs in Sudan and 231 lbs in Pakistan.

The failure to increase foodgrains production during the Third World Plan and resulting sharp increase in the food prices resulted in the development and enunciation of a new strategy, namely Intensive Agriculture District Programme (IADP) in a limited area. In 1964-65 a modified version of the approach was extended to several other parts of the country in the form of the Intensive Agriculture Area Programme (IAAP). A major change occurred with the introduction of the high yielding varieties and the programme was taken up on a full-fledged basis from 1966 onwards (Fourth Plan: pp. 114).

The High Yielding Varieties (HYV) programmer has so far been taken for wheat, paddy, *bajra*, maize and *jowar* the new strategy has yielded significant result and has stimulated many people to talk about success of the 'Green Revolution' and food self-sufficiency in early 1970s. The most striking success has been achieved in case of wheat and the production of maize, *bajra* and *jowar* has also shown slight improvement. But in the case of rice, the proper HYV seed has yet to be introduced on a large scale (Economic Survey, 1969-70, pp. 8).

Opinions about the performance are, however, conflicting. The most optimistic group holds the view that the so called green revolution is round the corner. On the other hand, doubts have been expressed both about the current performance and future progress of the project. The use of the package and practices does not yet seem to be common. The production of cultivators using the new seed is a little more than half even in wheat where the progress is great. The dosages of why a fertilizer used are usually much smaller than those recommended? Seed treatment is not as common as it should be. The use of pesticides and intercultural practices still has to go sufficient volume of credit and of facilities of irrigation are already acting as to be used only on irrigated lands which constitutes only small part of the total cultivated land in our country. Despite various limiting factors green revolution will increase the investible surplus both for the modernization of agriculture and the growth of non-farm sector (Sen, 1970).

Government of India decided to extend irrigation facilities for achieving higher agricultural output. Out of 545 major and medium irrigation projects undertaken since the beginning of First Five Year Plan, 325 projects have been completed and an additional irrigation potential of 24 million acres from major and medium irrigation projects has been created. The progress in extending irrigation facilities has always been behind the targets fixed in every plan. Several of the major projects were started under political pressure without complete investigation. In some cases, canal systems had been badly designed and farmers were not educated in advance for making use for the available water, there has been remarkable development in the use of groundwater resources because of the droughts of 1965-66 and 1966-67.

Lack of responsive leadership who could properly guide the farmers and provide necessary facilities for conducting agriculture on sound lines worked as an inhibiting factor in farm production. The Government, therefore, during the post-independence period decided to introduce Community Development Project (CDP) in which it was decided to encourage the efforts of the villagers to improve their economic, social and health conditions. The core aim of Community Project was to promote self-reliance, self-help and self-imitation among farmers and others who were living even in remote rural areas. In agricultural field the aim was to facilitate the process of diffusion of improved agricultural practices such as increased use of fertilizer and selective use of seeds, method of cultivation and the planting of products more suitable to local soil and climatic conditions. In short, Community Development Project has been one big institutional attempt to bring about a rapid transformation in rural areas in different walks of life.

However, because of various limitations CDP could not bring about the desired growth in the agricultural sector. It was, therefore, thought that a concerted and intensive programme, on a somewhat selective basis, was essential for creating the massive revolutionary change, necessary to bring about substantial growth in agricultural output. The programme is popularly known a package programme which aims at making an all-out efforts to raise agricultural output by providing the necessary incentives and was introduced on pilot basis in 1960-61 in the form of Intensive Agriculture District Programme (IADP). A modified version of the same approach in the form of Intensive Agricultural Area Programme (IAAP) was further extended in 1964-65 in different parts of the country. The programme has been successful in making a large number of farmers aware of the potentialities of the improved seeds, fertilizers, plant production measures and crop yields have also recorded substantial increase.

The Famine Commission (1945) had clearly visualized that organic manures (farm yard manures, compost, bone meal, etc.) would fall short of the potential requirements and had attached great importance to the use of artificial fertilizer for increasing yield. The Commission had further emphasized the application of fertilizers, about the correct proportion of nitrogen, phosphorous and potassium for different soils and crops and the proportion in which organic manure should be used along with fertilizers.

Green Revolution and its Implications

Green revolution is one of the biggest success stories of India citied globally, which enabled the country to convert the nightmarish 'begging bowl' status to that of 'self-sufficiency.'

Adoption of innovative techniques to usher in a technological revolution for the transformation of traditional agriculture into modern farming is designated by the happy phrase "Green Revolution". It was Willion S. Gaud, Former Director of the U.S. Agency of International Development (USAID), who was the first to use the term Green Revolution in a speech in March 1968 addressed to the society of International Development to describe significant changes in the agricultural sector in certain regions due to adoption of new farm technology. This new trend of agriculture is called more explicitly the seed fertilizer revolution as high yielding variety of seed is highly fertilizer intensive and capable of achieving a higher rate of yield (Dantwalla, 1970).

By 1960-61, stage had been almost reached where we could no longer think of producing more of farm products through extensive cultivation in order to meet the growing requirements of fast-increasing population. It has been well recognized that higher productivity is the only answer to meet the future requirements of food and other farm products. Social justice related issues such as equitable distribution of farm produce among cross-section of population, payment of wages to the agricultural labourers as per the inflation rate and justice to the share-croppers, etc. where not at all addressed by the preambles of green revolution. The improved variety of seed, no doubt, proved better in terms of yield. But it was not capable of achieving significant increase in the yield levels. It was recognized that the introduction of HYV seed in association with other modern inputs could only help to increase productivity and thereby meet the growing requirements of food and other farm products.

In one sense, the origin of Green Revolution could be traced back to the year 1960-61 when the Intensive Agricultural District Programme (IADP) was launched in seven districts in India. Initially pioneered by the Ford Foundation, the IADP stressed the need for providing the cultivator a complete package of practices consisting of new inputs, technical advice and credit. However, the improved variety of seed used could not achieve spectacular yield. It is generally believed that Green Revolution in India began in 1965-66 with the introduction of a new variety of maxican wheat.

The term "Revolution" implies a sudden change. But the change in agriculture in terms of application of crucial input like irrigation is not so sudden. Introduction of high yielding variety requires intensive irrigation. Since a long time, huge public investments have been made on major and medium irrigation works. Further, the HYV seed is confined mainly to food crops like wheat, paddy and *jowar*, etc.

The introduction of HYV seed led to many changes with several linkages. Irrigation became a prerequisite for the introduction of HYV and modern inputs. Therefore, the problem of water management now becomes important. Further, HYV seed is highly responsive to fertilizer and hence there is a remarkable rise in the demand for fertilizer. The HYV crop attracts pests and diseases, leading to demand for pesticides, sprayers etc. for cure. Attempts were made to address all these issues both at the institutional and at the individual level. However, at the individual level use of new facilities were highly selective. Affluent farmers used all these facilities in a disproportionate manner. The small and marginal farmers were pushed at the margin. Chaudhary (1999) underlined three major consequences of Green Revolution in Bihar. These are:

> 1. Due to cultivation of cash crops like tobacco, cotton, sugarcane, etc., the system of payment in cash became very common. But the minimum wages fixed by the government were never paid. Hence, whenever there was price rise, the cash wage was not capable to meet the need. This frustrated the agricultural workers. In the traditional system when wages were paid in kind there was no need to bother for the price rise of foodgrain by the workers.
> 2. The gap between the rural rich and the rural poor was widened by this revolution. Since farming became capital intensive, the big farmers invested in it huge quantity and earned a lot. The landless workers and the marginal farmers did not gain anything from the much publicized green revolution. This all frustrated the share-croppers, marginal farmers and the agricultural workers. Most of these people were from the low and the Scheduled Castes.
> 3. Although the traditional system of wage changed from kind to cash, the prosperous farmers preferred to grow cash crops instead of production for consumption, but they did not, however, change their behaviour towards the Scheduled Castes, which remained subject of abuse. Yadav and Kurmi landowners who themselves had suffered humiliation for long period, imitated the upper castes in their treatment of those who worked on their fields. Whereas, the benefits from agricultural modernization were narrowly held among the small class of rich and middle landowners, the anger against unfair economic treatment and social abuse spread widely among agricultural labourers. These agricultural labourers were mostly from the scheduled castes, and lower backward castes. (pp. 183)

Arora and Agrawal (1989) said that the green revolution brought self-sufficiency in foodgrains in India. The major contribution to this has been made by the Punjab state. As a result, the rural scene of Punjab witnessed many socio-economic changes. Infrastructure development like increase in the number of energized tube-wells, percentage of area irrigated, electrification of villages, road network, number of regulated markets, foodgrains storage etc. increased significantly.

The benefits of green revolution were, no doubt reaped by all sections of the rural society but differently. Farmers spent their increased income on improving their quality of life which is evident from the increased number of school going children, improving their housing structure and having sufficient amount of drinking water facilities.

Singh (1976) said that Eastern Uttar Pradesh has witnessed some spectacular change in farm front. The controversy that whether these changes have heralded a real green revolution or merely wheat revolution is only academic exercise. The facts remains that the introduction of new high yielding varieties of cereals has definitely raised the hope for a green revolution in the country, if it has not actually ushered it in. But it appears to be still far away from achieving our goals as is clearly evident from the situation created by droughts in a large part of the country during 1972-73. We have still to reach a stage when agriculture will start playing a significant role in developing the overall economy. This state of affair is mainly because the so called green revolution is confined to small and limited pockets of the country.

Green Revolution not only brought quantitative change in terms of increased production of foodgrains but also improved their quality of life. The deficiencies of proteins, calories, preservation taste appearance etc. were overcome by genetic improvement (including hybridization), supplementation of nutrients, control of plant disease etc. for instance, nitrogenous fertilizers improved the protein content in cereals.

People in rural area are becoming conscious of balanced and nutritious diet. They are more aware of the intake of calories, protein, fat and other nutrients. Customary patterns of food consumption of cereals, vegetables and other foods has helped in overcoming the nutritional diseases. Earlier the farmers could eat one vegetable grown in their fields for days together. One vegetable was grown only for home consumption, but now they can buy a variety of vegetables even from the village shop. Tea and beverages are replacing *lassi* and milk for serving the guest. Dress patterns has also undergone change in the rural area of Punjab. Middle aged men and older people usually use *Kurta* or shirt with "*tahmat*" (*lungi*) in the traditional dress. Women generally use *Salwar Kameej* of synthetic material with matching *Chunnies* (*Duptta*). Use of cosmetics among women have also increased (Gill & Saini, 1987).

Gangwar (1989) conducted a study on the impact of Green Revolution on consumption pattern and nutritional status in Haryana in 1970-71 and 1981-82. He concluded that on account of an increase in production and income of the cultivators, due to Green Revolution, per adult male unit expenditure on consumption has increased from Rs. 875 in 1970-71 to Rs. 2,274 in 1981-82. Green Revolution has also contributed towards an increase in the production and consumption expenditure and greater use of qualitative food, milk and milk products. In turn, it has improved the health and efficiency of farmers. This in turn has resulted in greater work and higher farm output and income.

Green Revolution and Food Security

The concept of food security is not new in India. In ancient India, it was visualized through diversification of farming based on physical and bio-diversity. The farmers grow all types of crops like cereals, pulses, oilseeds, millets, spices, fibres, vegetable and fruits. Mixed farming and mixed cropping was a tradition and almost all the farmers had cows, buffalos and bullocks and marginal farmers maintained goats and poultry. All these animals and crops helped in the recycling of animal waste and plant residues which helped in the enrichment of soil fertility through addition of organic manure. The Indus valley produced surplus food and payment to labourers was made in kind. All important cities had large storage facilities for stocking grains. The earlier rulers had also the wisdom of maintaining buffer-stocks to meet natural disasters and calamities. One granary, in those periods, stored enough barley to provide wages for 400 days while another granary had the capacity to pay in kind for 10,930 man-days. The priests, artisans, carpenters, blacksmiths and other functionaries received their wages in kind from the food producers under the *Jajmani* system. The concept of "Food for Work" was prevalent in those periods.

Food security is a product both of an adequate supply of food, which is production and easy access to it, which is purchasing capacity and system of distribution. The important considerations in maintaining long-term food security needs are production on sustainable basis, distribution through decentralized system and easy access to food through effective purchasing power. The World Bank has defined food security as 'access by all people at all times to enough food for an active, healthy life'. The essential elements of food security are availability of food and the ability to acquire it in a dignified manner. Definition of food security may be that every individual has the physical, economic and environmental access to a balanced diet and includes the necessary macro and micro nutrients, safe drinking water, sanitation, environmental hygiene, primary health care and education so as to lead a healthy and productive life.

Food security is normally identified through either nutritional or socio-economic indicators. Nutritional assessment may underestimate the need because it is a trailing indicator, which identifies only those who are already mal-nourished and not those who are poor or vulnerable. Mal-nutrition is the most unequivocal indicator of food security, as it is associated with a sudden and dramatic decline in food in-take.

The important reasons of large scale Famines in 1886 in the states of Bengal and Orissa were shortage of food and problems of quick transport and movement of food. The Famine Commission of India (1880), the Government of India Act of 1910 and the Royal Commission on Agriculture influenced the then British Government for speedy agrarian reforms, recognizing agriculture as the state subject and establishment of Indian Council of Agricultural Research (ICAR) as the apex body for research and development exclusively in the field of agriculture.

Another important policy effort aimed at increasing food production after Second World War turn economy was "Grow More

Food Campaign" (GMFC) in 1947. The report of Ford Foundation in 1959 resulted in the implementation of Intensive Agricultural District Programme (IADP). These were subsequently followed by Intensive Agricultural Area (IAA) and High Yielding Variety Programme (HYVP) in 1966-67 to attain self-sufficiency in food production by 1970-71. Another policy effort towards food security was the appointment of National Commission in Agriculture (NCA) in which projected production need and suggested measures to achieve them was envisaged. The latest thrust is now towards watershed development approach for rain-fed areas and Command Area Development Approach (CADA) for irrigated areas. All these efforts enabled the country to increase its food production from 54.92 m/t in 1949-50 to 198.96 m/t in 1998-99. The result was an increase in per capita consumption of food which improved from a low of 359g in 1951 to about 500g per day in 1997 despite the population increasing from 361 million in 1950 to 950 million in 1997..

There has been considerable impact on food production of the high yielding wheat and rice technologies. In 1964-65, there was probably less than 5,000 hectare area, in the developing countries, planted with the new high-yielding semi-dwarf wheat and rice varieties. By 1990 there was more than 130 million hectare area covered under these plant types.

Table 1.2: Estimated changes in wheat and rice production, yield, area and input use in developing countries, 1965-1994

Sl. No.	*Particular*	*1965*	*1975*	*1980*	*1985*	*1994*	*Percentage change 1965-94*
1.	Population (millions)	2210	2950	3280	3650	4350	97
2.	Wheat & rice production (million tonnes)	303	451	527	652	739	144
3.	Yield (tonnes/ha.)	1.5	1.8	2.3	2.8	3.2	113
4.	Area (million ha)	190	288	234	234	230	21
5.	Fertilizer (million Nutrient tonnes)	4	20	38	46	65	1525

Source: Agriculture Today, February 2001, p.11.

The new wheat and rice varieties have played a catalytic role for many other technological changes in agriculture in Asia. The use of inorganic fertilizers, which restores fertility to mined out soils, has increased more than 15-fold, the irrigated wheat and rice area has expanded by 50 per cent and the amount of irrigation water available for wheat and rice cultivation has more than doubled. The high yielding wheat and rice technologies also led to positive spillover effects on other crops, as farmers sought to improve overall productivity in other farm enterprises. Improved food production and food security greatly stimulated industrialization, which considerably increased employment. The so called green revolution in cereal production is an organic process as yield and production continue to increase.

One of the greatest contributions of Green Revolution has been the accomplishment of the goal of improved agricultural productivity while saving the additional land which would have been otherwise required to feed the population with traditional methods. A look at the figures makes it clear that if the same wheat and rice yields prevailed today that did in 1961-65, India would have needed an additional 94 million hectares of similar quality of land to produce wheat and rice.

Table 1.3: Profile of wheat and rice production in India and land saved through productivity gain

Year	*Area ('000 ha)*		*Yield (kg/ha)*		*Production (000 tonnes)*		*Land saved (000 ha)*	
	Wheat	*Rice*	*Wheat*	*Rice*	*Wheat*	*Rice*	*Wheat*	*Rice*
1970	16626	36360	1209	981	20093	35770	7582	NA
1975	18111	37890	1338	1045	24235	39580	11087	2487
1980	21962	38970	1437	1082	351560	42180	16061	4027
1985	23100	41000	1909	1418	44100	58300	30032	18329
1990	23500	42170	2120	1756	49850	74053	36560	33827
1995	25490	41640	2560	1952	65240	81260	53112	41194

Source: *Agriculture Today*: February, 2001.

The intensive agriculture project popularly known as 'Green Revolution' in which high levels of inputs like chemical fertilizers, irrigation, pesticides and energy are used was successful in increasing food production. Randhawa (1974) indicated that green revolution in Punjab has benefited all farmers—big, medium and small. It has also benefited farm labourers who command high wages in cash and also fed by the farmers. The phenomenal increase in production of foodgrains has not only saved the country from a grave political crisis, but also provided stimulus to the economy. A large demand for consumer goods has arisen. The requirements of the new agriculture, in terms of fertilizers, plant protection chemicals; pump sets, electric motors, sprayer and dusters are tremendous.

The intensive cultivation of maxican wheat taught the use of fertilizers to the farmers. Having learnt this new technology, they are applying it in other crops like sugarcanes, cotton, oilseeds, fruits and vegetables in a successful manner.

In the strategy of transforming traditional agriculture into modernized agriculture incentive to guide and reward farmers are a crucial component. The price support policy for wheat adopted by the Government of India is a key factor, which has assured the success of wheat revolution.

Swaminathan (1989) said that Green Revolution in Punjab is not miracle. It happened only because the following prerequisites for its success existed in the mid-sixties:

(a) Land consolidation and levelling,
(b) Owner cultivation resulting in a long-term stake in land,
(c) Rural communication,
(d) Rural electrification, and
(e) A dynamic Agricultural University.

Agricultural Modernization Today

Green revolution is one of the biggest success stories of India cited globally which enabled the country to convert the nightmarish "begging bowl" status to that of "self-sufficiency". It also brought about an element of resilience in agriculture towards off the threat of famines. The green revolution obviously ushered in an era of overall rural prosperity. Its impact was so dramatic that India became a role model for many developing nations.

The impressive transformation was possible due to major factors such as; (i) land reforms, (ii) creation of infrastructure such as irrigation, seed production and manufacture of fertilizers,

(iii) building of a strong agricultural R & D system, and (iv) Human resource development. In addition, the hard work of farmers was a deciding factor.

The resilience of agriculture was put to the acid test in 1987 when India successfully faced the worst drought of the century and more recently the sanctions imposed by some of the developed countries. Concerted efforts towards technology generation subsequently led to other landmarks in agricultural production such as the White Revolution (milk), Yellow Revolution (oilseeds) and Blue Revolution (fisheries).

Paroda (1999) said that the Indian Council of Agricultural Research (ICAR) has been in the forefront, guiding and coordinating these developments. As a result, the total foodgrains production increased from a mere 50.8 million tonnes during 1950-51 to 199 million tonnes in 1996-97. This was a four-fold increase while the population had increased three-fold. Wheat (68.68 million tonnes), rice (81.20 million tonnes) and oilseeds (25.6 million tonnes) have recorded an all time high production. All this has obviously been possible as a result of the adoption of good quality seeds, enhanced use of fertilizers and certified/quality seeds. Farm production increased to 6.5 lakh tonnes in 1994-95 from a mere two lakh tonnes during 1970-71, fruit, tropical, sub-tropical and arid, are cultivated 3.94 million hectares having a total production of 50 million tonnes, enabling India to surpass Brazil and occupy the first position in the world. In vegetable production, India is now the second largest after China with 72 million tonnes.

Singh (2002) also highlighted some facts on the agricultural modernization. To him, as a result of adoption of improved inputs and modern management practices, the total foodgrain production increased from a mere 50.8 million tonnes in 1950-51 to 208.87 million tonnes in 1999-2000 and productivity also increased from 522 kg/ha to more than 1697 kg/ha. The productivity of wheat, rice and oilseeds increased better than other crops.

The Indian farmers have adopted modern scientific agriculture. The high yielding varieties have been a catalyst in accelerating the pace of development in agriculture. Assured irrigation is the major requirement for adopting improved agricultural inputs and for increasing productivity. Besides, utilizing natural precipitation, the use of irrigation pumps has increased to more than 9.8 million electric pumps and 4.9 million diesel engine operated pumps to provide assured irrigation. The farmers are also using sprinkler, drip, micro-irrigation and greenhouse to increase the productivity and quality of produce.

Tractor was introduced through importation in the 1950s. Indigenous production started in 1961-62 with 880 numbers. Today the number of tractor in the country is estimated to be more than 2.63 millions. But the average availability of tractor is less than 2 per cent of the total farm holders. The use of mechanized agriculture can also be viewed through the increased use of diesel and electricity in the agriculture sector.

In this connection very interesting facts were highlighted during the collection of references that 'Deshi' ploughs, bakhar and patella were the most popular traditional implements for seedbed preparation prior to the 1960s[1].

Cultivators disc harrow, mould board plough, peddler, disc harrow-cum-puddler, peg tooth harrow, spring time harrow, rotavator and patella harrow operated by animal and tractor are improved machinery which have been adopted by a section of farmers. The line sowing not only saves seed but also facilitates regulated application of fertilizer near root zone. Besides, it helps control of weeds through the use of mechanical weeders. The animal drawn *Dufan* (two row), *Tifan* (three row) are used by the farmers as these cover more area and cost less. For precise application of seed and fertilizer, mechanically metered seed drill and seed-cum-fertilizer drill operated by animal and tractor have been developed and are being manufactured to suit specific crops and regions. Till drill and strip till drill have also been developed to reduce energy consumption. Inclined plate planter and pneumatic planter are the precise machinery for sowing; potato planter, groundnut planter, mechanical rice planter and sugarcane-set cutter-cum-planter are recent introduction in Indian agriculture.

Today in the area of agricultural modernization, intercultural and plant protection equipments such as sprayers and dusters are used by farmers. Sickle, *Khurpi* and spade are the major traditional hand tools for harvesting and digging. Self-sharpening serrated sickle has been developed with better material, which requires less frequent sharpening. Walk behind and self-propelled reaper harvesters are available which facilitate quick harvesting. Traditionally, bullock trampling, which is arduous and takes more time has achieved threshing. This has been gradually replaced by power threshers operated by 5-15 hp engine or electric motor. Tractor, PTO operated threshers are available for custom hiring. More than 2,50,000 power threshers are introduced every year. Combine harvesters are being used for harvesting wheat, paddy, soybean and gram in some of the states.

In short, the history of technological modernization in the field of agriculture is very old. However, during the post-independence period under different schemes technological modernization was introduced at the wider level on sustained basis by the Government of India in order to make the country self-sufficient in the production of food items. Farmers of different variety particularly big and medium, participated in this process. This led to economic development at the individual and community level which subsequently impacted inter-caste relationship, landowner-wage earner relationship, food habit, dress pattern, social status and position etc., in both functional and dysfunctional sense.

Krishi Vigyan Kendra (KVK) Intervention

Realizing the scope and importance of integrated working of interrelationship between research, education and extension, the ICAR established Division of Agricultural Extension. It was intended to enforce this functional relationship in research institutions, agricultural universities and allied institutions.

There are four main Transfer of Technology (TOT) projects of the ICAR going on at the national level. These are: (i) All India Coordinated Project on National Demonstrations (AICPND), (ii) Operational Research Project (ORP), (iii) Lab-to-Land Project (LLP) and (iv) Krishi Vigyan Kendra (KVK). All the projects of mobile type, except the KVK, are mainly the training institutions.

The projects are shifted every 5-6 years from one location to another in the large interest of the farming community and the extension worker of the area.

Training of farming and in-service extension functionaries is an input for the rapid transfer of agricultural technology. This aspect could not get due attention in the past, and thus it continued as a weak link to the agricultural production system. The existing training institutions were by and large theoretically oriented and hence they could not make much dent on increasing production income in the farming community as was expected. Realizing this crucial gap, the ICAR evolved KVK as an innovative institution for vocational training in agriculture and allied subjects.

According to Prasad, Chaudhary and Nayer (1987), Mohan Singh Mehta Committee (1974) was appointed by the Indian Council of Agricultural Research (ICAR) to formulate the KVK scheme. KVK is designed to impart skill oriented vocational training to the practicing farmers, increase field level extension workers, and to those who can ensure self-employment. Concept and Philosophy of KVK are the following:

(i) The Kendra will impart through work experience and hence, will be concerned with technical literacy, the acquisition of which does not necessarily require as a pre-condition to the ability to read and write.

(ii) The Kendra will impart training only to those extension workers who are already employed or to practicing farmers and fishermen in other words; the Kendra will cater to the needs of those who are already employed, or those who wish to be self-employed.

(iii) There will be no uniform syllabus for a Kendra. The syllabus and programme of each Kendra will be tailored according to the felt-needs, natural resources and the potentials for agricultural growth in that particular area. The three fundamental principal, viz. (i) agricultural production as the prime goal, (ii) work-experience as the main method of imparting training, and (iii) priority to working sections of the society, are the backbone of the

KVK programme. The social justice idea is to influence the productivity to achieve social-justice for most needy and deserving weaker sections, small and marginal farmers, agricultural labourers, drought and flood affected farmers and so on.

Need-based training courses are designed for different types of clients. Courses are based on the information received through family and village survey. No certificate or diploma is awarded irrespective of the duration of the courses. After the training, follow up extension programme is organized for converting the acquired skills of the trainees into practice. While designing the courses, the concept system is taken into account.

The training starts from farmers production units such as fields, farm, dairy units, poultry units, shop units, goods units, pig units, workshop, etc., and terminates in discussion assembly. The training programme takes into account all methods and means, which may rest in skill development in trainees in the areas of their interest. It can be informal, and non-formal or a combination of all these depending upon the needs and resource of the farmers.

The main objective of the KVK is to provide a strong training support for brining about production break through in agriculture. The specific objectives of KVK are as below:

(i) Plan and conduct survey of the operational area to prepare the resource inventory with special reference to identify the training needs of the farming community.

(ii) Compile all relevant recommendation/package of practices for the district to be meaning utilized in the training courses and the follow up extension programmes.

(iii) Plan and conduct production-oriented, need based, short and long duration training courses both on the campus as well as in the village for various target groups with priority on the weaker and poorer sections.

(iv) Organize farm science clubs both in rural schools and in village to inculcate in the younger generation a liking for and an interest in agriculture and allied sciences and for scientific farming through supervised projects. Develop and maintain the campus farms and demonstration unit on scientific lines as the facilities for providing work experience to trainees as also dissemination of the latest technical know-how.

(v) Providing practical facilities of the Kendra to the teachers and the students of vocational agriculture of higher secondary schools.

(vi) Impart some general education to the rural illiterates and school drop-outs in order to make them not only good farmers but also better citizens.

(vii) Provide training facilities in home making and nutrition education for rural community and gradually encompass other important area such as home crafts and cottage industries, consistent with the requirements of integrated rural development in collaboration with concerned organization, the society like the tribal farmers, and marginal farmers, agricultural labourers, drought and flood affected farmers and so on.

There are relatively generalized institutional objective of the Kendra. The specific objective, nevertheless, may be many and the varying depending upon the agro-ecosystem for increasing the economic status of the targeted groups.

Thus, each Kendra was developed with its specific objectives geared to the immediate as well as long-term needs of the operational area.

Conformation with the innovative character of the Kendra is not an easy task. Concerted and continuous efforts have to be made to realize the ideals of the Kendra by way of imparting tailor-made skill training. Kendra made action plan on the basis of following features:

(1) The training courses, in order to be relevant, should be based on a deliberate attempt to identify the training needs which is complied with the help of a scientific survey, group interviews and scientists visit/observation, etc. The survey

being a time consuming task, the help of sister institutions such as Agriculture University, Research Institute, etc. may be sought.

(2) Long-term and short-term training courses are organized on campus or out of campus.

(3) In order to impart skill training by providing work experience, it is essential to develop ideal training (facilities such as farms of various sizes consistent to the farm size of the area), farm workshop, livestock demonstration units, poultry units, goat units etc. The trainer/scientists and KVK intimately involved is responsible for developing and maintaining these facilities.

(4) Each KVK unit prepares its own syllabus or Action Plan on the basis of local requirements and the specific needs of the targeted groups.

(5) The KVK is composite in nature both in respect of subject matter and the audience (trainees). In other words, the training facilities have to be developed in all the subject matters/areas/district, and the trainees are both male and female, boys and girls, old and young, literates and illiterates. This is in the demand of inter-disciplinary approach and multi-institution linkage and support.

(6) In order to be effective, the Kendra develops and maintain functional linkages with allied institutions.

(7) A local management committee with representatives from the agricultural university, ICAR, relevant government departments, progressive farmers, etc., is formed for each KVK for providing guidance on its working.

(8) The achievement of KVK is measured in terms of training courses organized, number of beneficiaries, follow up field extension activities organized, and the impact of these activities on the farming community in general and agricultural production in particular in social, economic, psychological and cultural sense.

Krishi Vigyan Kendra (KVK) at the National Level

The KVK project is sponsored by the ICAR and it is implemented by the ICAR institutes, agricultural universities, reputed voluntary organizations and state department of agriculture. The first KVK was established in 1974 at Pondichery under Tamil Nadu Agricultural University. Upto December 2012, total 631 KVKs have been established throughout the country by all agencies.

The priority for the establishment of KVK is given to hill areas, drought-prone areas, forest areas, costal areas, flood prone areas, and areas dominated by tribal farmers, weaker sections, small farmers and landless labourers. This is primarily in order to promote inclusive and sustainable development of agriculture by giving special emphasis on disturbed areas as well as the marginalized population.

In Fifth Five-Year Plan total 19 KVKs were established, and during Sixth Five Years Plan total 70, and during Seventh Five Year Plan 20 and during 1991-92 total 74 KVKs have been established in the country. In Eighth Five-Year Plan total 78 KVKs have been established.

In Ninth Five-Year Plan ICAR had decided that total 53 Zonal Agricultural Research Stations (ZARSs) will be converted to Krishi Vigyan Kendras, so that these stations will work on the KVK mandates at district level. In Ninth Five Year Plan total 66 new KVKs were approved. Upto 2002, 286 KVKs and 53 ZARSs have been well established and working according to ICAR.

According to old data total 153 KVKs are working under the State Universities, 79 KVKs are handled by Non-Governmental Organizations (NGOs), 31 KVKs are working under the ICAR and remaining KVKs are working under the Education Institutions, Public administration etc. Maximum 32 KVKs were approved in Uttar Pradesh and 31 in Rajasthan.

In the Eighth Five Year Plan total Rs. 160 crore budget allotment was approved while in the Ninth Five Year Plan total Rs. 102 crore were proposed.

During the Ninth Five Year Plan total 55 thousand training programmes were organized. Total 1.3 million farmer's, landless labourers and farmwomen were benefited from this programme. About 2.5 lakh rural youth and 1.4 lakh extension personnel were benefited in their training programme.

During the Ninth Five Year Plan KVKs had organized total 52 thousand extension activities, i.e. *Kisan Mela, Kisan Sangosthi,* Field day, Radio and TV talks, Film shows, Exhibitions, etc

In the year 2002 total twenty-six thousand demonstrations were conducted on farmers field under front line demonstration programme on oilseed, pulses and cereals. This demonstration became one of the factors for an increase in the production of cereals up to 3587 tonnes, 681 tonnes of oilseed and 507 tonnes of pulse (Annual Report of KVKs, 2002, ICAR, New Delhi). Hence, since its inception an attempt was made by the Government of India to increase the number of KVKs at the national level to meet its objectives.

KVK in Madhya Pradesh

In M.P. also an attempt was made to establish KVKs in all the districts. It is needless to mention that the state is not only inhabited by a large number of SC and ST population but also many of its districts are rain-deficit with having poor quality land and soil.

By the end of 2007 altogether 47 KVK's were working in different districts. Of these 8 were established by NGOs and government departments and remaining 39 were established by Jawaharlal Nehru Krishi Vishwavidyalaya, Jabalpur. KVK Chhindwara was also established by JNKVV, Jabalpur. District-wise details of KVKs given in Table 1.4.

KVK in Chhindwara

The KVK Chhindwara was established in 1983 under Jawaharlal Nehru Krishi Vishwavidyalaya (JNKVV), Jabalpur. The KVK Chhindwara could organize following activities in the district during the period of 1983-84 to 2000-2001.

Training were organized in the mentioned period on the crop production (555), Horticulture (202), Animal Husbandry (184), and Agro-forestry 43. Agro-forestry discipline was covered in the year of 1995-96. Total 72025 beneficiaries like farmers, farm women and rural youth participated in these trainings.

Table 1.4: Krishi Vigyan Kendras in Madhya Pradesh

Sl. No.	*Address of Krishi Vigyan Kendras*	*Host Organisation*	*Year of Sanction*
1.	Krishi Vigyan Kendra, Village Hamirpur, Distt. Datia.	Vice Chancellor, Jawahalal Nehru Krishi Vishwavidyalaya, Jabalpur	2007 SAU
2.	Krishi Vigyan Kendra, Village Aawari Distt. Ashoknagar.	Do.	2007 SAU
3.	Krishi Vigyan Kendra, Village Shankarpura Khurd and Sandas Khurd, Distt. Burhanpur.	President, Lokmata Devi Ahilya Bai Holkar Social National Mission, Masjid Compex, Shanwara, Burhanpur-450331, M.P.	2007 NGO
4.	Krishi Vigyan Kendra, Village Piparsama Distt. Neemuch	Vice Chancellor, Jawaharlal Nehru Krishi Vishwavidyalaya, Jabalpur	2006 SAU
5.	Krishi Vigyan Kendra, C/o S.K.Tripathi House No.207, Near Subha Motor, Katla, Distt. Mandla	Do.	2006 SAU
6.	Krishi Vigyan Kendra, Bataggakhurd, Distt. Barwani	Do.	2006 SAU
7.	Krishi Vigyan Kendra, Village-Darbrohatrict Distt. Umaria	Do.	2006 SAU
8.	Krishi Vigyan Kendra, MP State Seeds and Farm Development Corporation, VIII. Baroda, Distt. Sheopur	Do.	2006 SAU
9.	Krishi Vigyan Kendra, Balgarh Farm, PO. Balgarh, Distt. Dewas, M.P.	Do.	2005 SAU
10.	Krishi Vigyan Kendra, PO. Katangi Kala, Village Padaria, Distt. katni.	Do.	2005 SAU
11.	Krishi Vigyan Kendra, Nowgaon, Pan Research Centre, Distt. Chhatarpur-472001.	Do.	2005 SAU
12.	Krishi Vigyan Kendra, Fatehpur, Distt. Shivpur	Do.	2005 SAU
13.	Krishi Vigyan Kendra, Powarkheda, Distt. Hoshangabad-461110.	Do.	2004 SAU

(Contd.)

Table 1.4 (*Contd.*)

Sl. No.	*Address of Krishi Vigyan Kendras*	*Host Organisation*	*Year of Sanction*
14.	Krishi Vigyan Kendra, PO. Jaura Khurd, AB Road, Distt. Murena-476001.	Do.	2004 SAU
15.	Krishi Vigyan Kendra, Zonal Agriculture Research Station, Bamhori, PO. Rajauna, Distt. Sagar-470002.	Do.	2004 SAU
16.	Krishi Vigyan Kendra, Zonal Agricultural Research Station Distt. Khargone-451001.	Do.	2004 SAU
17.	Krishi Vigyan Kendra, Girwar Farm, Distt. Shajapur-465001	Do.	2004 SAU
18.	Krishi Vigyan Kendra, Near Vikram Nagar Railway Station, PO. Box No. 202, M.L. Nagar, Distt. Ujjain-456010.	Do.	2004 SAU
19.	Krishi Vigyan Kendra, College of Horticulture, Distt. Mandsaur-458001.	Do.	2004 SAU
20.	Krishi Vigyan Kendra, Distt. Jabalpur-482004.	Do.	2004 SAU
21.	Krishi Vigyan Kendra, Distt. Harda.	Do.	2004 SAU
22.	Krishi Vigyan Kendra, Jaiswal Quarters, Civil Ward No.2, Distt- Damoh-470661.	Do.	2004 SAU
23.	Krishi Vigyan Kendra, Shastri Bhawan, Station Ganj Area, Distt. Narsinghpur.	Do.	2004 SAU
24.	Bhoj Krishi Vigyan Kendra, Near Village Naktara, PO. Bankhedi, NH-86 Ext, Raisen Sagar Road, Distt-Raisen-466551	Secretary, PDKVAAS, Bhopal	2004 NGO
25.	Krishi Vigyan Kendra, Tribal Agricultural Research Station, Distt. Dindori-481880.	Vice Chancellor, Jawaharlal Nehru Krishi Vishwavidyalaya, Jabalpur-482004	2003 SAU
26.	Krishi Vigyan Kendra, College of Agriculture, Distt-Gwalior-474002.	Do.	2003 SAU
27.	Krishi Vigyan Kendra, Kuthulia Farm Distt. Rewa- 486001	Do.	2003 SAU

(*Contd.*)

Table 1.4 (*Contd.*)

Sl. No.	*Address of Krishi Vigyan Kendras*	*Host Organisation*	*Year of Sanction*
28.	Krishi Vigyan Kendra, Betul Bazar, Distt. Betul-460004.	Do.	2002 SAU
29.	Krishi Vigyan Kendra, Laxmipur, Distt. Panna-488001	Do.	2002 SAU
30.	Krishi Vigyan Kendra, Post Box No.18, Distt. Dhar-454001.	Do.	2002 SAU
31.	Krishi Vigyan Kendra, CRDE, Vill-Sewaniya Ichhawar, Distt. Sehore-462043.	The Chairman, Centre for Rural Development and Environment, Bhopal	1999 NGO
32.	Krishi Vigyan Kendra, Etawah Road, Distt. Bhind-477001	Vice Chancellor, Jawaharlal Nehru Krishi Vishwavidyalaya, Jabalpur-482004	1997 SAU
33.	Krishi Vigyan Kendra, Kasturba Gandhi National Memorial Trust, Kasturbagram, Khandwa Road, Distt. Indore-452020	The Chairman, Kasturba Gandhi National Memorial Trust, Kasturbagram, Indore.	1996 NGO
34.	Krishi Vigyan Kendra, College of Agriculture, Jaswadi Road, Distt. Khandwa-450001	Vice Chancellor, Jawaharlal Nehru Krishi Vishwavidyalaya, Jabalpur-482004	1994 SAU
35.	Krishi Vigyan Kendra, Kundeshwar, Distt. Tikamgarh-472001	Do.	1994 SAU
36.	Krishi Vigyan Kendra, Block Office Compound, Distt. Seoni-480661	Do.	1994 SAU
37.	Krishi Vigyan Kendra, Kothi Bagh, Biaora, Distt. Rajgarh-465661	Do.	1994 SAU
38.	Krishi Vigyan Kendra, Raghogarh Naka Aron Distt. Guna-473101	Do.	1994 SAU
39.	Krishi Vigyan Kendra, Shiksha Samiti, Kalukheda, Distt. Ratlam-457339	President, Kalukheda Shiksha Samiti, Ratlam.	1994 NGO
40.	Krishi Vigyan Kendra, Karaundia, Distt. Sidhi-486661	Vice Chancellor, Jawaharlal Nehru Krishi Vishwavidyalaya, Jabalpur-482004	1992 SAU
41.	Krishi Vigyan Kendra, Technical School Campus, Rewa Road, Distt. Shahdol-484001	Do.	1992 SAU

(*Contd.*)

Table 1.4 (*Contd.*)

Sl. No.	*Address of Krishi Vigyan Kendras*	*Host Organisation*	*Year of Sanction*
42.	Krishi Vigyan Kendra, Badgaon, PO. Pala Distt. Balaghat-481115	Do.	1992 SAU
43.	Krishi Vigyan Kendra, Sri Malwa Mahila Vikas Samiti, PO. Sirjon, Distt-Vidisha-464228	President, Malwa Mahila Vikas Samiti, 32, Niyamatpura, Shahjehanabad, Bhopal	1992 NGO
44.	Krishi Vigyan Kendra, Majhgaon, Distt. Satna-485331	Chairman, Deen Dayal Research Institute, Rani Jhansi Road, New Delhi.	1992 NGO
45.	Krishi Vigyan Kendra, JNKVV Farm, Rajgarh Naka, Distt. Jhabua-457661	Do.	1984 SAU
46.	Krishi Vigyan Kendra, Chandangaon, Distt. Chhindwara-480001	Do.	1983 SAU
47.	Krishi Vigyan Kendra, Berasia Road, Distt. Bhopal-462038.	Director, CIAE, (ICAR), Nabibag, Berasia Road, Bhopal-462038.	1979 ICAR

Source: Obtained from www.icar.org.in dated 15/04/2013.

Table 1.5: Activities of KVK Chhindwara during 1983-84 to 2000-01

Sl. No.	*Activates*	*Number of Conduction*
1.	Training Courses	1816
2.	Training Beneficiaries	58108
3.	Demonstration	2736
4.	No. of Beneficiaries	2736
5.	Total No. of village covered	363
6.	No. of Extension activities like field day, Sanghosti, Exhibition, Film show, Radio talk, Kisan Mela etc.	682
7.	No. of beneficiaries from Extension activities	72025

Source: Status Report of KVK, Chhindwara (1983-84 to 2000-01).

KVK Chhindwara conducted demonstrations on the farmers' fields in different situations of the district. Total 2736 demonstrations were organized under the programme like National Demonstrations (ND) and Lab-to-Land Programmes (LLP). In the year 1990-91 the ICAR started a new mandate, i.e. Front Line Demonstrations

(FLDs) on oilseeds and pulses. KVK Chhindwara also conducted FLDs on oilseeds—Soybean, Linseed, Niger and Sunflower, while FLDs on pulse—Gram and Pigeonpea (*Arhar*) was organised. In the meantime, a demonstration programme was also launched by ICAR that is front line demonstration other than pulses and oilseeds. This programme was organised on the basis of local needs and local crops, vegetables, poultry, dairy, agro-forestry and plant protection managements and women in agriculture. KVK Chhindwara also conducted FLDs on kitchen gardening, poultry keeping, dairy, plant protection/management, agro-silviculture system under agro-forestry, vegetables—cauliflower, onion, potato and pea on the adopted farmers' fields in the adopted villages.

KVK Chhindwara is working on the basis of ICAR guidelines. Firstly, KVK scientists collected local facts and local needs through survey, workshop and Participating Rural Appraisal (PRA), then they prepared technological intervention modules with the help of subject specialists suiting to local conditions. After preparation the Action Plan of KVK, a Zonal workshop is organized every year and each KVK makes presentation of its own action plan. According to Zonal Coordinating Unit Zone VI (ICAR) headquarters at JNKVV building, Jabalpur action plans are approved and after completion of the progarmme, an annual progress report is also presented at the time of Zonal Workshop. Each KVKs submit monthly, quarterly and half yearly progress reports to their host institute (JNKVV) and ZC unit. The monitoring teams from ICAR, New Delhi also visits KVK, farmers, fields, demonstration sites and interact with the beneficiaries. KVK Chhindwara has well maintained KVK due to which it got an outstanding first KVK award during 2001 from ICAR, New Delhi.

Researchers have been visiting KVK Chhindwara to see progress of KVK. There are 128 extension literature in Hindi which were developed and published in different new specific technologies, total 392 newspaper coverage, 122 radio talks released on local need based technologies, 05 TV talks were also released during 1983-2001 period, total 147 popular articles were published in different magazines of India level, state level and as well as 71 research papers were published in the journals and magazines. KVK also prepared some special publications like Technical Bulletin, Pictorial book of agricultural engineering, impact study of Transfer of Technology programme by KVK in Badnoor village, and pictorial presentation of woman participation in training programme. A decade of National Demonstration in Chhindwara district, progress report on demonstration and popularization of improved implements and machinery through KVK 1991-92 to 1996-97, Photo montages, success stories and farmers interviews for ready references are prepared for monitoring team.

Under front line demonstrations the following varieties of crops were demonstrated on the farmers' fields.

Table 1.6: Crop-wise varieties demonstrated

Sl. No.	*Crop*	*Varieties*	*Character*
	Kharif		
1.	Arhar	ICPL-87 (Asha)	Short duration & high yielding
2.	Soybean	JS 71-05	High Yielding
		JS 335	High Yielding
		JS 1024	Rust tolerant
3.	Groundnut	GG-2, Gangapuri	High Yielding
4.	Castor	GCH-4	High Yielding
	Rabi		
5.	Linseed	Kiran, JL 23-0	High Yielding
6.	Gram	JG 74	High Yielding
	Wheat	JG 174	Short duration
	Vegetables		
7.	Cauliflower	Snowall-k	High Yielding
8.	Pea	Arkile	High Yielding
9.	Onion	N-53	Agri found dark red

Source: Status Report of KVK Chhindwara, 2001.

KVK in Simariya, Jaam and Chargaon Villages

The people of Simariya, Jaam and Chargoan Karbal are mainly dependent on agriculture. But their traditional agriculture could not satisfy their ever-increasing requirements. They got very nominal yield from their land holding. Farmers had obtained low yield, due to lack of upgraded seed, old and outdated varieties, more infestation of insect and pest, less and unbalanced use of fertilizer, unknown about use of micronutrients, weedicides and insecticides. Farmers are growing kharif, rabi and vegetables since many year and have experienced substantial losses due to occurrence of diseases, infestation of insects and pests, under doses of fertilizer and upgraded seed. Most of the farmers do not pay attention to the damage by disease, due to lack of knowledge about suitable resistance variety.

Almost the entire area of soybean was under improved varieties with high production potential, while the yield were found low. The major problems identified were imbalance use of fertilizer, more infestation of insect, improved selection of varieties in light medium soil, high intensity of weed, and poor quality of seed and attack of rust. The farmers belonging to small production system category, faced the problem of money for purchase of proper varieties and good quality of seed etc.

Wheat was cultivated after soybean, *Jowar*, maize, and vegetables hence, generally sown late. The yield of wheat was very low in spite of the fact that farmers grew high yielding varieties. The major causes of low yield, as identified, were improper plant population, imbalance use of fertilizer, poor quality of seed and heavy infestation of *Bathus* weed.

Gram crop in this area encounters damage due to gram pod borer and poor plant population. The pest incidence declines the gram yield considerably. Farmers do not know about selection of pesticides, its doses and time of application. They also do not observe the economic threshold level.

Potato is the vegetable crop in the selected villages. Generally, potato crop was cultivated after harvesting of soybean crop. The use of poor quality seed was a common practice in the selected villages. Farmers used to go to the market and purchase seed from the vegetable shopkeepers. Actually, the good and big small potato were available for planting purpose. The major cause of low yield of potato, as identified, were infestation of diseases, plants having less number of shoots and poor plant growth.

Bhindi was grown as vegetable in Jaam village. The low yield of *Bhindi* during rainy season was due to high infestation of insect, delay in adoption of plant protection, attack of virus disease, imbalance and low dose of NPK, improper selection of varieties and non-availability of resistant varieties.

Dairy was traditional enterprise in all the three villages and on an average each family had at least one or two milch animals along with bullock pair. Green fodder was the most important component of feed for such livestock for supplementation of protein and vitamins. There was always a competition of green fodder with the food crops, land being the limiting factor and agriculture being the first most important source of livelihood.

To overcome pathetic situation of farmers and to uplift their socio-economic standard, an effort has been made by KVK through their mandates as a change agent.

So far as introduction of KVK in the sampled villages are concerned, the project entered in these villages during 1994-95. Although KVK surveyed and contacted 10-15 villages of Mohkhed Block of Chhindwara district but these three villages could successfully achieve the target, which was designed by KVK.

Social and economic conditions of these three villages are by and large similar to the neighbouring villages but it was the high density of population, larger population size, dependence of majority of villagers on land, relatively better agricultural activities and interest among the farmers to produce vegetables as a cash crop were some of the broad factors, which invited KVK to initiate its operation in these three villages. Numerical dominance of weaker sections, location of villages on road and near to the KVK centre, and covered mass spread area like Chhindwara to Nagpur road, (Simariya), Umera nalla to Betul road (Jaam and Chargaon Karbal), ambitious response of villagers and quick support building between the villagers and the KVK officials were also factors which led to selection of all three villages.

According to KVK, quick establishment of farmers club on the initiative of KVK by the villagers worked as enabling factor. At the time of Transfer of Technologies Programme very needful technologies were transferred and farmers of these three villages easily adopted and availed good achievements. One very strong point also raised at the time of our study was that farmers from these three villages were regularly contacted and expressed their views as a feedback. KVK scientists were assessed and refinement in technology was made. In the year 1995-96, Participatory Rural

Appraisal (PRA) was also conducted in these villages. Some young and energetic farmers were chosen as a key informant, they regularly visited all fields of villagers and collected information about actual needs and valuable intervention, which helped farmers to adopt KVK initiatives easily. There was no conflict among the farmers, not casteism and not favouritism. KVK conducted almost all activities in these villages, like trainings, demonstrations, On Farm Testing (OFT), Kisan Sanghoshti and farmers and field days. Farmers of three selected villages showed their views at the time of study in a groups "If any team comes for monitoring of KVK work then it must be brought to their fields for visiting", this was said by one of the villagers. He further said "Quinnuennial Review Team (QRT) (ICAR) came to visit our villages on 14 June, 2001. Total three hours monitoring and evaluation at Jaam village with farmers and one hour in Simariya village was made. While villagers of Chargaon Karbal participated, at Jaam. QRT members raised only one query whether whole activities of KVK are conducted only in Jaam village?

On the basis of above analysis one can easily see positive relationship between technological modernization in agriculture and high farm production and subsequently high income. Green Revolution is an example. KVK may also be perceived as an extension of Green Revolution and its philosophy. But how and to what extent KVK led economic development has impacted socio-cultural life and practices are the issues which need documentation. Present study is an effort in this direction. The study is both exploratory and descriptive in nature.

NOTE

1. Availability and use of traditional farm technology is rampart even today in the tribal areas. For instance, in a study (Chaudhary, 2012) it was observed in a tribal village of Madhya Pradesh that of the 201 households 101 had plough, 98 had bakhar, 18 had Kulpha and 44 had bullock cart. In contrary to this 1, 4, and 16 households had thresher, sprayer and pump set respectively nobody had tractor. However, some villagers used to hire tractor from neighbouring villages in need (p. 52).

CHAPTER – TWO

Earlier Studies

Trends of research on interface between agricultural modernization and social change is an old process. It was extensively conducted during 50s and 60s when Community Development Project along with other initiatives for rural development were undertaken. However, majority of these studies in the due course of time dealt with economic impact of such interventions. Majority of these studies concentrated on the examination of rural development, community development and technological changes. The present chapter neither reviews a large number of studies on interface between agriculture modernization, economic development and social change, nor it attempts to review the related literature from all the states where the KVK was introduced. Purpose of this chapter is to find out related and relevant issues covered in the past and identify research gap so that the present study may be conducted in a specific direction. Some of the studies reviewed, however, are as under.

Haque & Khan (1999) conducted study on the impact of Barind Multiple Development Project on socio-economic development in Rajshahi region of Bangladesh. Study was conducted in 1997 to examine the differences between the pre-project situation and the current situation of the Barind project in Bangladesh in terms of cropping intensity, irrigation facilities, irrigated command areas, yield of crops, wage rate, family income, etc. Data were collected from 175 respondents, including Zonal and Area officers in Rajshahi, Noagaon and Nowabgaon district as well as project beneficiaries and general farmers in and around the Barind Multiple Development Project. It was found that a significant increase in deep tube wells took place in the project area and fallow land was converted to cultivable land to a great extent, single cropped area and indigenous method of irrigation decreased and double and triple cropped area increased. Cropping intensity, irrigated command area and family income also increased by a significant amount.

Frankel (1971) while making visit to Ludhiana (Punjab), West Godavari (Andhra Pradesh), Tanjore (Tamil Nadu), Palghat (Kerala), and Burdwan district (West Bengal) where Integrated Agriculture Development Programme (IADP) was introduced, came to the conclusion that as a result of the introduction of HYV seeds and other modern inputs, the yield per hectare had gone up and consequently, almost all categories of farm producers were benefited. In Ludhiana district, where HYV wheat was grown widely, farmers were able to gain substantially. But it is stated that the additional gains from new farm technology were not properly distributed among different social groups. Farmers with 5 to 10 acres of land were able to increase their net income to some extent, but those with 15 to 20 acres of land growing wheat were able to make adequate investment on the new inputs and achieved substantial gains. It was also noticed that in rice-growing regions, even though small farmers were able to gain to some extent, their real income did not improve because of the considerable increase in the production costs. In the matter of yield per hectare of land, the large size farmers did not make any headway as compared to small farmers. It was also noted that in the rice growing regions, the economic conditions of about 75 to 80 per cent of the farmers had deteriorated. Also, majority of the landless population were left away from this modernization process. Hence, Frankel concluded that agricultural modernization was both gainful as well as painful.

Sachchidananda (1972) studied several IADP areas in Sahabad district of South Bihar. His study revealed that the higher caste people with more education, higher socio-economic status and larger holdings had maximum adoption of improved practices, seeds, fertilizers etc. The medium size family showed the largest adoptions. The effect of education of the head of the household on adoption behaviour is felt only after reaching the middle school standard. Those farmers in whose family there are members serving in towns adopted faster than others. Head of households in the age-group of 20 to 34 led in adoptions. The face of young farmers turned definitely towards change. The standard of living was slightly higher in the IADP areas than in the non-experimental Blocks. This is noticeable in the larger use of articles of luxury in such areas. There is no appreciable difference in the level of aspirations of farmers in the IADP areas from those in areas outside it. The concept of an ever-rising level of aspirations does not obtain in the areas under study. The lower the level of living, the more static are the aspirations. Sometimes, the peasant does not consider it moral to want more. There is little attempt to innovate or adopt an innovation. If more income is generated from agricultural development, it may be wasted in non-essential forms of consumption, instead of being invested to maximize production. There is a tendency to treat the small farmers as inefficient cultivations. But the fact is that they are being squeezed out of their holdings. There is an attempt by the landowner to evict sharecroppers forcibly in order to take up direct cultivation or cultivation through hired labour. The rural gentry and the rich peasants have consolidated their economic and political position. Agriculture development and the resultant prosperity have not made much difference in the matter of observance of caste norms. The process of modernization has not made any impact on any social institutions. The inter-caste social distance is almost the same as before. Caste is still the governing principle of social behaviour. Friends, relations, neighbours and caste leaders are still the main channels of communication for introduction of innovations. Technological change no doubt increases the income of all types of farmers, but increase in income is not uniform among the farmers. Rich farmers are greatly benefited because they can afford to purchase modern inputs in adequate quantities, while the poor farmers, due to lack of finance, cannot take the best advantage of the modern inputs in time[1].

Hence, like Frankel, Sachchidananda is also of the view that technological modernization is both gainful and painful for farmers and villagers depending on their socio-economic status.

Randhawa (1974) studied the impact of Green Revolution in Punjab. To him, Punjab is different from other states of India, as the villagers are more educated. There are about half a million serving soldiers and pensioners in the countryside. Apart from these there are large number of teachers, minor Government officials, policemen, and truck, taxi and lorry drivers who have roots in the countryside. All these skilled men have played a productive role in agriculture. They have invested savings in land development, and have also intelligence and education to deal with the intricacies of tractors, machines and plant protection chemicals. In fact, the best agriculture is practiced by family members some of whom are also educated. Among the totally illiterate, agriculture is very poorly managed. The Green Revolution in Punjab has benefited all farmers, i.e. big, medium and small. It has also benefited farm labourers who command high wages in cash and also fed by farmers. The phenomenal increase in production of foodgrains has not only saved the country from a grave food crisis situation, but also it has provided stimulus to the economy. A larger demand for consumer goods has arisen. The requirement of the new agriculture, in terms of fertilizers, plant protection, chemicals, pump sets, electric motors, sprayers and dusters are tremendous. The intensive cultivation of Maxican wheat taught the use of fertilizer to the farmers. Increased grain production has given a stimulus to poultry farming. People whose diet was mostly vegetarian, now consume a protein rich diet. In the strategy of the transforming traditional agriculture into modernized agriculture, incentives to guide and reward farmers are a crucial component. The price support policy for wheat adopted by the Government of India is a key factor, which has assured the success of Wheat Revolution. Hence, Randhawa strongly highlighted functional consequences of green revolution related to economy and physical quality of life. But socio-cultural consequences of green revolution was seldom underlined by him.

Bhattacharya (1976) in the study of farmers of Narayangarh block in Midnapore district of West Bengal, observed that farmers in this block are hesitant and very cautious in accepting the new improved seeds and techniques of cultivation. When new inputs came to them, they were reluctant to accept them. But as a result of persistent endeavour by the Block Administration and Village Level Workers (VLW) and the examples set by progressive farmers, they are gradually shaking off their inhibitions and traditional prejudices. Once farmers are convinced of the efficacy of an innovation, they do not offer resistance to it out of their blind love for traditions. It is, of-course, true that their extreme caution and concomitant resistance have slowed down the process of diffusion of modern farm practices but this caution is bred by the fact that the farmers in general always look for security in raising crops, because failure is not simply a material loss to them but loss of their livelihood for a full year, the adverse effect of which have often to be borne out by them in successive years also. The farmers do not, therefore, usually want to add a further element of uncertainty to an undertaking which is already fraught with plenty of uncertainties. Hence, the need for constant persuasion was strongly advocated by him in this matter.

Singh (1976) conducted a case study in eastern Uttar Pradesh on modernization of agriculture. The results clearly showed that due to modernization of agriculture, increase in total production in this area has resulted in the supply of foodgrain to *mandi* in Varanasi. In 1964-65 one of the wholesale dealers in Varanasi, who hails from this area, constructed big godown in this Block at Bajranagar to supply grains for sale in the area. But the surplus production forced him to convert it into a workshop and he has started manufacturing thresher and doing the repairs of irrigation and other farm machinery. As a result of bigger marketed surplus and net savings, farmers are investing the additional resources on farm machinery, new tube wells and construction of better houses for themselves. This has also resulted in the creation of a new class of transport operators in the area. An approach road is visually non-existent; camels are used to transport farm surpluses to the roads and farm inputs from the roads to the interior villages. New brick-kilns have now started coming up in the area which supply bricks for pump houses and for the construction of *Pucca* residential buildings. As already stated, their engineering workshops have been established in the area and these are doing good business. Hence, from sustainable development point of view technological modernization has done excellent job because farmers have invested their surplus income in trade and commerce which will improve and retain their improved status even when they withdraw themselves from farm work.

Dasgupta and Madan (1978) studied two villages of Uttar Pradesh. According to them, both these villages presented two contrasting profiles of community structure. Structural features also characterized Ladpur village, which had a higher level of agricultural development than Pampur village, which was characterized by its heterogeneity in its caste structure. It had a relatively high degree of integration as a community. Both land ownership and leadership were held by the members of the dominant agricultural castes, which had the skill, capital, time and interest to invest in agriculture, which in turn contributed to high degree of agricultural development. The leadership pattern in this village was differentiated by agricultural leadership and was reputed as being general leaders of the community. The leaders had enough influence on farmers to induce them to utilize the resources effectively. Thus, as a community, Ladpur village was in the process of modernization and development. Such a community structure was related positively not only to agriculture development but also to other kinds of technological change and development. Pampur was homogeneous in terms of caste and Yadavs were dominant in the village both in numerical and economic terms. Although Yadavs were in large numbers and they owned the major proportion of the available land, but their farms were relatively small in size. Yadavs also lacked modern farming skills, capital, time and interest to invest in agriculture. Cultivation to the Ahirs was low subsistent activity and one of the multiple sources of income. The intermediate and Scheduled Castes in the village owned very small-sized farms, the productivity of which were very low. Agriculture was not a differentiated specialized activity in Pampur. Consequently, there was no agricultural leadership in Pampur. Pampur also lacked a well-recognized leadership structure although one person was almost universally recognized as a leader because of his prestige and kinship status in the dominant caste.

In other words, the leadership in Pampur was informal and traditional in orientation. Pampur lacked any differentiation of interest areas, and consequently, there was no generalized leadership in the village. Pampur's community structure was also segmented. The intermediate and Scheduled Castes in the village had very little opportunity to participate in the affairs of the village. As a result, the members of the intermediate and Scheduled Castes expressed feelings of apathy and negative conviction regarding the potential of their community for development and the effectiveness of its leaders. The rigidity of the community structure in Pampur was manifested by the fact that not a single person belonging to the intermediate or Scheduled Castes was mentioned as a leader by the villagers. The community structure in Pampur was not very conducive to agricultural or other kinds of technological development.

Thekkamalai (1983) conducted study in Pithalaipatti village of Tamil Nadu. The modern socio-economic forces have its impact on Pithalaipatti social life, as on that of any other Indian village. The general changes in the village economy like the change from subsistence to cash economy, has taken place in Pithalaipatti also. The farmers now concentrate on commercial crops like sugarcane and cotton and produce them for outside market. Changes have taken place in the techniques of agricultural production. Agricultural innovations are widely adopted chiefly because of the efforts of community development personnel. Although changes have taken place on the production side, there is less change on the marketing side. The Co-operative Bank offers farmers facilities for selling their agricultural products, but most of them depend on the commission agents of Dindigul town for selling their agriculture produce. The hereditary economic relations of the village have weakened. Only a limited number of households engage the village artisans and the service caste people on traditional payment. Even the artisans and service caste people like the Vannan prefer cash payment for their services. Their patrons also prefer cash payments for account of the high price of grain. The introduction of pumpsets for irrigation has affected the work of cobblers and ironsmiths. The cobblers have lost their traditional work of stitching the leather for the *moat* for lifting water from the irrigation well and the ironsmiths that of making the drum to attach the leather and providing iron shoes for the bullocks. Most of them now work as agricultural labourers and work as sandal repairers.

The processes of diversification of occupations in different fields have now taken place. Now there are mill workers, small traders, teachers, clerks, constables, postmasters, gram-sevak and so forth and so on from Pithalaipatti. People give importance to formal education for securing salaried jobs. There is not much change in the ritual and religious relations between different caste groups. The matures and the Prayers beat the drums at the funeral ceremonies of the caste Hindus. The Vannan and the Barbar continue their ritual services. Though the Scheduled Caste groups want to raise their social status, they are not able to challenge the higher caste villagers merely with the help of social legislations. The Scheduled Castes of the village are mostly landless agricultural labourers and depend mainly on mercy of the high caste landowning groups for work. They cannot dream in the near future of sowing against the upper castes. The study also showed that caste barriers have been weakened in most of the social interactions. The feeling of social distance is negligible in social situations which are less intimate in nature, and it is pronounced in situations with deeper emotional involvement. The feeling of social distance steadily increase in situations where intimacy is greater. The feeling of social distance is less among caste groups of equal status and it increases with the lower status of the other caste groups. The villagers have slowly started building their houses close to the S.C. colony, but still they leave some space between their house and houses of SCs. The SCs are allowed to use the same cups in the village tea stalls, but they are not treated as equals in the homes of the higher caste groups. The belief in pollution through physical contact is not strictly held. Only old and conservative people follow it, all young people irrespective of their caste disregard this practice. However, in festivals the same old custom of caste differentiation is followed.

Sharma and Diwakar (1989) conducted study in two villages of Palampur, Himachal Pradesh under the "Lab-to-Land Programme". The complete list of households with their landholding was obtained and small farmers were selected. These families were visited by a team of scientists almost every week from 1982-83 to 1983-84, to educate them about the available technology and its application. Result of the study have disproved strongly held myth that the small farmers are fatalistic in nature and they lag behind in exploiting the opportunities available to them to maximize their production. The results have proved that there exists a potential on small farmers which constitute as high as 70 per cent of the total holding in Indian agriculture. The potential could be exploited by strengthening the existing extension facilities through *ad hoc* progrmme like Lab-to-Land. The study also showed that innovations once introduced are bound to be adopted due to the demonstration effect on the experimented farmers. The results of the study also distinctly suggest that even if the present subsidies on fertilizers and other inputs are withdrawn these are not going to adversely effect the production.

Chaudhary (1990) conducted a study in select villages of Madhya Pradesh. The scope of the study was to examine impact of biogas technology on woman's health and time management system. He reported that majority of respondents experienced great relief from drudgery of smoke and fuel borne disease. The study also revealed that majority of respondents felt free from disease related to eyes and cough. The study also concluded that few numbers of respondents were not suffering from eyes ailment and water was not flowing from their eyes. However, due to the introduction of new cooking technology there was no saving of time from cooking work in the significant manner. This is primarily because now some new varieties of food were cooked. While in the past guests and visitors were served with *gur* and a glass of water but today they are served tea. Many farmers today consume more than one vegetable at a time. Also, the adopter families felt enhancement in their social status. It was also observed that the experimental households now demand more dowry because they claim that they will provide more comfort to their would-be daughter-in-law. Hence, impact of biogas technology may be perceived both positively and negatively.

Mishra (1992) presented a synthetic analysis of various aspects of agriculture sector ranging from land use pattern and infrastructure to the cropping intensity, production and productivity in Mirzapur district of eastern Uttar Pradesh. He attempted to assess the impact of growing agriculture sector on the development of rural areas.

He concluded that the progress made during recent years in the field of agriculture have made a significant contributions to the process of rural transformation and change in the area. The changes in agriculture sector have direct bearing on the process of rural development. The innovative and commercial agriculture produces positive impact and helps farmers to accelerate the process of rural development, while the primitive and subsistence agriculture reverses the direction. Thus, rural development is the natural corollary of agricultural development. It can be safely inferred from the study that agriculture sector in Mirzapur district is now gradually but steadily moving from subsistence stage to the modern commercial stage. The increased productivity and production of various crops have brought about tangible changes in income and living standard of rural people.

This has been made possible due to following reasons, viz.

(1) Compulsions of growing demands of growing population for qualitative as well as quantitative changes in agriculture production,
(2) Gradually expanding network of information facilities,
(3) Favourable attitudinal changes in the farming community towards the use of inputs and innovative methods in agriculture,
(4) Growing awareness about multiple cash cropping,
(5) Growing awareness about modern agriculture,
(6) Growing knowledge about demand and supply,
(7) Increasing demand for agriculture raw materials, and
(8) Increasing consciousness of the peasant community about the role of agriculture in the generation of income and assets and formation of capital.

Because of the impact of above factors agriculture has made appreciable improvement and inturn, has brought about positive changes in several areas of rural sector. The important ones are:

(1) The rise in per capita income and assets of the farmers,
(2) Improvement in the living standard of villagers,

(3) Increased calorie intake by the people, low incidence of deficiency disease and decreasing mortality rates,
(4) Increasing literacy and education in rural areas,
(5) Improving housing conditions and gradual emergence of *pucca* houses almost in all villages,
(6) Emerging integrated spatial pattern of rural settlement,
(7) Growing hierarchical order in the spatial organization of central functions and service, and
(8) Expanding base of rural internalisation in the district. It can, therefore, be concluded that agriculture sector growing under a variety of environmental, social and economic constraints, has contributed a lot to the process of rural transformation and change in Mirzapur district of eastern Uttar Pradesh.

Narayanamoorthy (1992) pointed out that the use of modern input and machineries like tractors, threshers, sprayers, manual and power pump sets, oil as well as electricity has increased rapidly after the introduction of new technology, but the intensity of the use of these modern machineries depends upon the nature of inputs, liquidity position of farmers and awareness of farmers about the modern machineries.

Singh (1993) conducted study on the impact of KVK on socio-economic and infrastructure utilization of farmers of Keonjhar district of Orissa. He observed that, in order to speed up agricultural production and economic standard of rural people, several special programmes have been put into operation. The Indian Council of Agricultural Research introduced KVK with an objective to provide a strong training support for bringing about production breakthrough in agriculture and allied sectors. Krishi Vigyan Kendra, Keonjhar started functioning during the year 1982. The tribal people of Keonjhar mainly depend on forest products. Their dependence on agriculture is insignificant. Gradual deforestation could not satisfy their hunger. They get very nominal yield from their eroded rainfed land. To overcome pathetic situation of tribal farmers and to uplift their socio-economic standard, the KVK made its efforts in this direction. It was significant to note that, after implementation of KVK Keonjhar, the income level of the respondents went up to

Rs. 3500 and Rs. 3500–7500 respectively. In case of non-beneficiaries group the respondents having income up to Rs. 3500 had been reduced from 50 to 20 per cent whereas second and third groups had been increased from 24 to 52 per cent and 8 to 18 per cent, respectively. Though, there was little change in income level, but significant change had not been occurred in this respect. In respect of extent of irrigation the beneficiary farmers (76%) had partially irrigated their land against 30 per cent found before the implementation of KVK. There was no significant change in land holding size of farmers. With regard to family size and its type, there had been significant change in family type of the farmers. There was an increase in family size. Household having more than 10 members were increased 16 per cent in case of beneficiaries and 20 per cent in case of non-beneficiary farmers. It was also observed that there was little change in social participation of both categories of farmers. There was little difference between beneficiaries and non-beneficiaries with regard to their cosmopoliteness after the implementation of KVK. The beneficiaries were more cosmopolite than the non-beneficiary farmers due to orientation with KVK personnel's from time to time. Regarding source of information, before the implementation of KVK the VLWs and AEOs were the main sources of information for both categories of respondents but after implementation of KVK the extent of information provided by the KVK personnel's was the main source for the beneficiary farmers. The non-beneficiaries received information regarding modern technologies from the beneficiary farmers, friends and neighbours.

Taneja (1994) focused on output employment linkage in agriculture in the context of new technology. It has also delineated various causative factors as accelerators or retarders of these two important variables. The finding leads to conclude that in view of the high positive effect of biochemical element of the new technology IFS (Irrigation, Fertilizers, Seed) on output and employment both, it is suggested that a socially desirable pattern agricultural development involving a high growth rate of output and employment is attainable only through the introduction of better seed qualities. This requires an active institutional and state support to research in biosciences, which could be possible through enhanced budgetary allocations to these items. Further, the indispensable nature of agriculture mechanization (tractorisation) due to its relatively higher positive contributions in the output then the bullock power, and its net negative impact on human labours employment leads to the conclusion that this harmful effect of mechanization could be mitigated only through large increase in farm produce, which as indicated by the results of the study, is crucially dependent on the IFS.

Improvement in the quality of land through the use of modern scientific techniques could be another variable to increase productivity in agricultural sector, which would obviously promote labour use. It may, however, be noted that there is certainly a limit to the extent to which employment opportunities in the farm sector could be increased through upward shifts in production function. This is particularly so in view of accompanying mechanization. It is, therefore, suggested that the policy should aim at creating more employment opportunities in the off-farm sector of the economy in order to seek solution to the malady of rural poverty.

Panda (1996) conducted study in Bhubaneshwar district of Orissa. Objectives of the study was to examine how efficiently various farm inputs are used by different tenurial farms in agriculture. The findings of the study show wide difference in the cropping pattern as well as crop yield. The study conclusively indicates limited impact on resource use and crop productivity. Hence, steps may be taken to provide security of tenures and other institutional facilities to the tenant farmers to have better use of resources and higher crop yield in agriculture.

Sharma (1996) conducted study on KVK Chhindwara (MP) on factors associated with knowledge and adoption of vegetable production technology by beneficiaries and non-beneficiaries of the KVK. He observed that there was higher knowledge and adoption scores of beneficiaries as compared to non-beneficiaries. Family size and farm power showed positive and significant relationship with beneficiaries. Their level of knowledge, education, farm credit, social participation, communication sources, farm literature and perception about change had significant positive correlation. Also, with reference to non-beneficiaries there was negative impact on occupation, farm size, income, type of house and farm power.

Jain & Mishra (1996) conducted study in four villages; two of them were located on roadside near to the city, each with a population of approximately 2000 persons, with having facility of electricity, a post office, and a high school. The other two villages were located in remote areas, at least 20 km away from the main road. These villages were without electricity but a post office, and education facility up to middle classes were available. The analysis of data revealed that the meaning of technology in the true technical sense was largely absent from the cognitive scheme of common men. In the light of their experience with agriculture technological inputs, the villagers could express certain specific and general attributes of technology. These respondents extended the word Science to include technology. Villagers indicated their experience and realization that the scientific and technological innovations form a new and salient force influencing their lines across all the important domains. It was found that the villagers had greater awareness of agricultural equipment than by domestic and recreational ones. Villages on road showed greater awareness than the remote ones. Finally, the study revealed that villagers are psychologically ready to accept the technology particularly in agricultural field but its availability is the main factor. At present only big farmers, who have sufficient land, are in a position to mobilize their resources to use technology. The remote and pooı villagers are still waiting. Villagers are quite certain about the positive and negative impact of technology. The negative impacts are by and large of social nature which seems obvious, particularly in Indian society, where agriculture is embedded into a hierarchical social system. The influx of technological change on the one hand ensure economic uplift but simultaneously disorganizes the traditional social system, whether it is for good or bad cannot be assessed at this transition period. It appears, at present, that technological input is conducive to raise the economic status of only rich farmers and thereby increasing the gap between poor and rich class. It seems to be a vicious circle that technological inputs need resources, which are only available with rich farmers and these inputs further enhances their resources and subsequently their socio-economic status.

Mohanty (1996) examined the impact of modernization patterns in the agrarian economy of the rural setting. In course of presenting his observations in relation to various technological and allied developmental activities in a particular village of Orissa, the author has pin-pointed how mechanization in agricultural activities brought prosperity to the region so far as the nature of land and the life of the people are concerned. But at the same time there are contrary effects. The small and poor peasants have been failed to associate themselves with new technology, new situation has become incompatible to them due to certain obvious reasons. In other words, rich and progressive farmers are availing better opportunities and they are drawing conspicuous benefits whereas the poor peasants and agricultural labourers have been facing economic insecurity which is further increasing and also likely to increase in future.

Khan (1997) conducted study on Tharu tribe of Uttar Pradesh. Tharu tribe is a settled farming tribe. A large number of Tharu farmers are covered under Integrated Tribal Development Programme (ITDP) in order to bring upliftment in their socio-economic conditions. For more than a decades Tharus have been exposed to new agricultural technology and its adoption has revolutionized their traditional farming and has brought about many changes in their economic behaviour. Thrust of his research was to study these changes and to compare the economic life of Tharus prior to 1983 and during 1993 that is before and after adoption of New Agricultural Technology (NAT). The results showed that after adoption of the NAT the yield level of rice and wheat had increased three times. There occurred significant change in the farming practices of the Tharus. The use of several improved devices indicated that farm mechanization was coming up very fast. In recent years significant changes can be seen in the total annual cropped area occupied by different crops. Instead of single cropping, now double cropping pattern has become a general practice among the Tharus and in some areas even triple cropping pattern is also noticed. Cropping intensity, market behaviour and storage pattern have also significantly changed during the post-experimental phase.

Rajput and Verma (1998) conducted study on the flow of institutional credit in agriculture in Indore district of Madhya

Pradesh. It was found that the average intensity of cropping on borrowers farm was higher at 140 per cent as against 122 per cent on non-borrowers farms. The input, output, new income per hectare of small, medium and large farms were significantly higher on the farms of borrowers than on the non-borrowers' farms.

Mangat and Gill (1999) conducted study to examine the impact of specialized training courses on beneficiaries in generating/ strengthening of self-employment avenues and augmentation of their income from various subsidiary occupations in Punjab. The first Krishi Vigyan Kendra (KVK) at Gurdaspur established by the Punjab Agricultural University, Ludhiana. The result of study clearly shows that the specialized training courses at the KVK created employment and helped to augment income of the trainees of all the courses. It also created employment for other family members and rural folks who can render their services either as casual labourers or permanent labourers. It can be concluded that the training in the above-cited subsidiary occupations has helped farmers to create self-employment and employment for other villagers along with consequent enhancement in their income.

Mohanty (2000) observed that agricultural modernization began systematically in a village of Orissa state after 1956 when water from Hirakud dam started flowing into the Bargarh canal. At present more than 77 per cent of the total cropped area was cultivated by tractor. There were 5 households in the village having one tractor each and 11 households having 16 pump sets. The mechanical thresher was also found among the upper landowning households. Sprayer was seen almost with all well-to-do peasants. The gross cropped area of the village constituted 380.83 acres. The HYV programme was initiated in the village in 1964 and today the area under HYV amounts to more than 87 per cent of cropped area. The per acre fertilizer consumption of the village worked out to 82.90 kg. The study also pointed out that upper castes were essentially the upper class and they acquired large quantity of land in course of time and mostly the households of lower castes and tribes joined the landless and marginal peasant categories on a large scale. Study also revealed that most of the landholders of upper castes exchanged their land according to suitability of canal irrigation with those of the lower castes and tribes. Author also observed that agricultural modernization through an introduction of new cropping pattern, water management system and labour relations, has increased the number of disputes among the peasants in the village. As the power structure of the village revolved around landowners who belonged to upper castes, in most of the cases decision of disputes went in their favour.

Thakur (2000) conducted study in Mandsour district of Madhya Pradesh in order to examine the effect of "Lab-to-Land Progaramme" on adoption of improved new technology, increase in productivity levels of field crops and financial impact on farmers. Most of the households belonged to backward and Scheduled Castes. Subsidy to the extent of Rs. 500 per family per year was made available to implement the programme which was supplied in the form of critical inputs, i.e. fertilizers, insecticides, pesticides, improved seed and so on. The finding reveals that there was significant change in cropping pattern and local varieties were replaced by high yielding varieties of seeds. The farm families became more conscious about seed treatment, plant nutrition and plant protection measures. There was tremendous change in fertilizer use on crops leading to higher yields of major crops grown by the adopted marginal and small farm families in the study.

Tharan's (2000) study shows that modern technology contributes significantly to an increase in rice and household income if it is associated with the development of infrastructure facilities, irrigation, flood control and drainage, transport and communication that results in efficient marketing. Study also indicated positive impact of technology and infrastructure on overall income distribution.

Chaudhary (2002) said that agricultural development led to the employment of many people who directly relied on it for their livelihood. In Rajasthan, for example, agriculture provides livelihood to 68.8 per cent of the labour force. Agricultural development also leads to the increased production of raw materials for industries which in turn created employment opportunity, revenue to the state and manifest and latent benefits. Finally, to some extent, agricultural development checks migration of worker to urban area and also leads to balanced interface between agriculture and industry.

On the basis of review of some of the relevant studies, presented above, following points are obvious. In most of the studies the role and importance of new technology in farm production were highlighted. Most of the researchers have underlined economic impact of technological modernization in the functional sense. Scholars have also highlighted increasing disparity between rich and poor due to an introduction of new technology. To them new farm technology has definitely contributed to self-dependence in the field of food but definitely it has taken social cost. Researchers have also discussed impact of inhibiting and enabling factors in this context. Role and importance of both input supply including capacity building were highlighted. But only few of the scholars have discussed the impact of economic development on social structure, social institutions and social practices. In other words, social impact of technological modernization is largely missing from the studies discussed above. Hence, the present study is an attempt to fill up this gap. More specifically this study intends to examine interface between technological modernization in agriculture, its economic impact, particularly from income point of view and subsequently on social institutions and practices like food habit and dress pattern, health and disease management, inter-personal relations, law and order, dowry system, untouchability, migration, education, small family norm, wage rate, gender discrimination, leadership pattern and so on.

NOTE

1. In this connection B. Sen (1974) opined that technological change has sustained improvement in the per capita income and standard of living of farmers who have been experimenting with it. Since the benefits of new agricultural technology are correlated to irrigation facilities, the size of holding, financial position, education, tenurial and social status of the farmers, the impacts of new agricultural technology on income have not been uniform in different regions and even among different types and size of farms within the same region.

CHAPTER – THREE

Methodology

Broad objectives of the study was to examine level of adoption, functioning and impact of Krishi Vigyan Kendra (KVK) on farmers. It also aimed to explore the nature and direction of social change in the light of KVK intervention. However, some specific objectives of the study were the following:

1. To study the functioning of KVK and popularization of its philosophy among cross-section of farmers in order to modernize farm activities;
2. To identify the factors causing uneven spread of the technology if any, amongst different social groups;
3. To find out the impact of agriculture modernization on economic development in-terms of annual income, saving, indebtedness, physical quality of life and assets creation; and
4. To find out process of social change in the villages under study in-terms of change in social structure, status of women and other components of non-material culture which might have taken place primarily due to the introduction of KVK technology.

Universe

Profile of the District

The study was conducted in Chhindwara district of Madhya

Pradesh. The district is located in southern part of the state of Madhya Pradesh. Geographically the district is situated between latitude 21°23' and 22°44' N and longitude 18°18' and 79°24'E. It is surrounded by Seoni district in east, Betul in west, Narsinghpur in north and Nagpur (Maharashtra) in south. Total geographical area of the district is 11815 sq. km. Among the nine agro-climatic zones of the state, viz. Malwa, Vindhyan, Narmada, Satpura, Jhabua, Gird, Kaymore, Bundelkhand and Nimar, the district of Chhindwara falls under Satpura region. There are 11 Development Blocks in the district. Location of all these 11 Development Blocks are like this.

Table 3.1: Sub-region-wise location of blocks in Satpura region

Sl. No.	*Sub-region*	*Blocks*	*Situated*
1.	Satpura Hills	Tamia and Harrai	North part of the district
2.	Satpura Plateau	Chhindwara, Amarwara, Parasia, Junardeo, Mokhed and Chourai	Central part of the district
3.	Satpura Plains	Sauser, Bichhua and Pandhurna	Southern part of the district

Source: Zonal Agriculture Research Station, JNKVV; Chandangaon, Chhindwara (MP).

The total cultivated land is 602098 hectare, which comprises 495929 monocropping and 119589 hectare double cropping area respectively. According to the land utilization record from Deputy Director Agriculture in Chhindwara (DDA), net irrigated area is 21.7 per cent and agriculture intensity is 124 per cent. The average rainfall ranges in between 1000-1617 mm. The average land holding size in the district is 2.33 hectare. The major crops of the district are Soybean, Maize, Jowar, Groundnut, Pigeon pea (Arhar), and Niger in Kharif and Wheat, Gram, Linseed, Sunflower, Pea, Potato and vegetables in the Rabi season.

According to census 2001, population of Chhindwara district is 1848882. Caste-wise population is like this: 12.2 per cent Scheduled Castes and 44 per cent Scheduled Tribes. The district has a total 11 Janpad Panchayats while only 9 Tehsils are in the district. There are 1903 villages in the district. The density of population is 156 per sq.kms.

Fig. 3.1: Map of Madhya Pradesh

Studied Villages

The study was conducted in three villages named Jaam, Simariya and Chargaon Karbal. All these villages are from Mohkhed Block of Chhindwara district of Central Madhya Pradesh. Profile of all the three villages under study in terms of agriculture, social and economic features etc. are as under.

Criteria for Selection of Villages

The Krishi Vigyan Kendra (KVK) was established under the Jawaharlal Nehru Krishi Vishwa Vidyalaya (JNKVV), Jabalpur in the year 1983, in Chandangaon village of Chhindwara district. It

Fig. 3.2: Map of Chhindwara District

was established to promote various types of Transfer of Technology (TOT) programme, i.e. training of farmers, training of women and rural youth, conducting of Front Line Demonstrations (FLDs) at farmers, fields on different crops, in-service training for extension personnel and organization of extension activities like, *Kisan Mela*, Field Days, Exhibitions, Film Show, *Kisan Sangoshti*, Radio talks, TV talks, Newspaper coverage and literature developed on different specific techniques. On Farm Testing (OFT) technique was also demonstrated on farmers' fields.

According to status report of KVK (1983-84 to 2000-01), total

1816 training programmes were conducted which consisted of 58108 beneficiaries. Total 2736 field demonstrations were conducted on different crops in different villages. Total 363 villages were covered during this period. KVK report also reported that maximum TOT programmes were organized in adopted villages of Mohkhed block. Mohkhed block have 182 villages and total geographical area of the block is 68989 hectare. Size of the total cropped area is 46794 hectare.

The main crops of the Block is Soybean, Maize, Jowar, Groundnut, Pigeon pea, and Niger in Kharif and Wheat, Gram, Linseed, Potato, Cauliflower, Pea and other vegetables like Tomato, Brinjal, Chilly, Garlic, Carrot etc. in Rabi.

Demographic Situation

According to 2001 Census Report following are the socio-demographic and physical features of all the studied three villages.

Rainfall and Climate

The sampled villages are situated within a radius of 10 km. Rainfall and Climate condition is the same. The villages have a sub-tropical climate with annual rainfall ranging from 1000-1300 mm with occasional winter showers. The average number of rainy days are only 51. Mostly the rainfall is received between middle of June to middle of September. The maximum rainfall is received in the monsoon period, i.e. between June to August. A very little rain is available in winter months, i.e. December and January. Rains are highly erratic which markedly affects agricultural operations. Except in the rain season, other months are comparatively dry.

The mean temperature varies from less than 8°C in winter to more than 45°C during summer, lowest temperature are observed in January and highest in May.

Soil

In general, the nature of soil of the sampled villages are light to medium belonging to the order of vertical. In local language its

Table 3.2: Detailed informations about the villages under study

Sl. No.	*Village information*	*Jaam*	*Simariya*	*Chargaon Karbal*
I.	Population of Village	2359	733	1800
	Male	1178	379	925
	Female	1181	354	875
II.	Number of Households	354	195	250
III.	Literacy			
	Male			
	Literate	1178	302	860
	Illiterate	—	77	65
	Female			
	Literate	1181	256	615
	Illiterate	—	98	260
IV.	**Land Utilization**			
	Geographical Area	605.284	416.907	389.257
	Cultivated area	510.00	372.00	360.00
	Irrigated area	111.00	65.399	68.550
	Double Crop area	136.392	65.110	141.253
	Cropping intensity	134%	148%	132%
V.	Education facilities	Higher Sec.	Primary	Middle & High
	Girls enrolled	136	5	87
	Child enrolled	153	9	29
VI.	Distance from District	25 km.	22 km.	26 km.
VII.	Transport/Road facilities	Yes	Yes	Yes
VIII.	Post Office	0 km.	3 km. Umranalla	2 km. Rajegaon
IX.	Railway Station	3 km. Umranalla	3 km. Umranalla	10 km. Umranalla
X.	Bus Stand	3 km. Umranalla	3 km. Umranalla	10 km. Umranalla
XI.	Block Headquarter	6 km. Mohkhed	5 km. Mohkhed	6 km. Mohkhed
XII.	Medical facilities	Primary Chhindwara Nagpur 100 km.	Primary Chhindwara Nagpur 100 km.	Primary Chhindwara Nagpur 110 km.
XIII.	Veterinary facilities	Yes	Yes	Yes
XIV.	**Irrigation facilities**			
	Well	354	103	188
	Tubewell	1	2	2
XV.	**Agriculture Equipment**			
	Tractor	2	2	5
	Electric Pumps	225	97	188
	Sprayers	5	8	25
	Duster	3	3	—
	Power thresher	1	2	11
	Sugarcane crasher	—	—	1
	Electric Crasher	—	—	1
XVI.	**Wages rates**			
	Male	40 Rs. /Day	40 Rs. /Day	40 Rs. /Day
	Female	25 Rs./ Day	25 Rs./ Day	25 Rs./ Day

Sources: Census Report, GOI, 2001 and personal observation.

name is *Barra* (medium) and *Mud Barra* (light). There is low availability of *Nitrogen* and low to medium availability of *phosphorus*.

Source of Irrigation

The major sources of irrigation are wells, stop dam, dam and tube-wells. On the basis of panchayat records there are 354,103 and 188 private wells in Jaam, Simariya and Chargaon Karbal respectively. Similarly, there are 9 tube-wells in Jaam, 10 in Simariya and 6 in Chargaon Karbal.

In the Jaam village 40 farmers lifted water from *Umariya Dalel Dam* (Basin area) through $2\frac{1}{2}$ to 3 inches underground pipelines. Temporary light connection is issued from MPEB, for the period between September to March. Generally, farmers get water through 1-3 km long pipeline, the cost of which comes to 1 to 3 lakhs.

Cropping Pattern

Cropping pattern is the same in all the three studied villages. Soybean, Maize, Groundnut, Jowar and Arhar are the main crops in the *Kharif* season. In *Rabi* season Gram, Linseed, Potato, Garlic, Carrot and vegetables are the main crops. Those farmers who have bigger landholding size and proper input facility today they prefer to produce more for market than their counterparts who have small landholding size and insufficient quantity of infrastructural facilities.

Marketing Facilities

The sampled villages face lack of marketing facilities. Umeranalla was the main place for marketing within 3 km. away from Jaam and Simariya. Farmers of Chargaon Karbal sell their produce to Mohkhed, which is about 6 km. away from this village. All the inhabitants of these three villages carry their agricultural produce to Chhindwara town. The vegetable like Garlic, Cauliflower, Pea, and Carrot, etc., are, however, transported to Nagpur by Matador/TATA 407.

Education

Jaam is the only place which has one Higher Secondary school. It is well-maintained with having good teaching staff. It is a lucky village because here horticulture subject is also taught at Higher Secondary level. There is co-education in the school. Girls are equally encouraged to continue education.

In Simariya there are two private schools and one government school. All these schools are upto primary level. Students of Simariya have to go to Umeranalla for education. They use bicycle or go on foot to reach Umeranalla. In Chargaon village there is a government high school and a private middle school. They also have to go to Umeranalla for further studies.

Health Service

Health care services have been recognized as an essential part of any human society. Better health care facilities tend to lower the mortality rate and prolong the life span of the individuals.

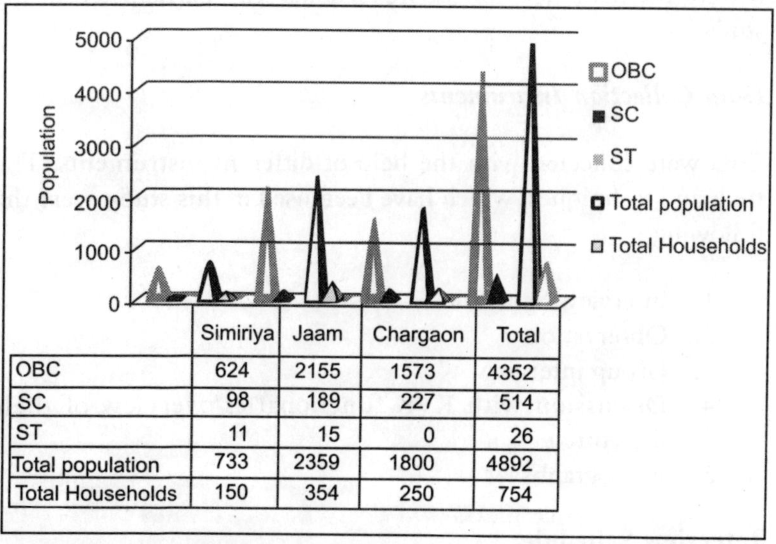

	Simiriya	Jaam	Chargaon	Total
OBC	624	2155	1573	4352
SC	98	189	227	514
ST	11	15	0	26
Total population	733	2359	1800	4892
Total Households	150	354	250	754

Fig. 3.3: Caste and Community-wise distribution of population/households

There is one primary health centre available with a staff nurse, 5 beds to cater to the health care need of inhabitants of all the three studied villages.

Caste

A wonderful data came out in the study. Nobody belongs to general caste or upper caste in the sampled villages, and SC population is residing in these villages out of which the OBCs are in majority. As far as the ST population is concerned it is very nominal, they resides in Jaam and Simariya. There is no ST resident in Chargaon.

Selection of Respondents

There are total 754 households in all the three villages, i.e. 150 in Simariya, 354 in Jaam and 250 in Chargaon. All the 754 households participated in the study. Hence, census technique was used to collect related and relevant information from households. Mainly it were the male head of the households who participated in the study.

Data Collection Instruments

Data were collected with the help of different instruments. The tools and techniques, which have been used in this study, were the following:

1. Interview Schedule
2. Observation
3. Group interview
4. Discussion with KVK functionaries/overview of their literature
5. Photographs

Interview Schedule

The primary data was obtained with the help of semi-structured interview schedule. Interview schedule was prepared on the basis of objectives of the study. All-important issues related to study were covered in semi-structured interview schedule. Before finalization of the instrument pre-testing of the instrument was also made. Some reliable and suitable corrections were incorporated and the schedule was finalized in the course of investigation; several visits were made from time to time to collect the information keeping in view the convenience of cultivators. Attempts were made to explain to the respondents, the purpose of the study so that the researcher could develop a rapport with the respondents. This technique helped in collection of data in a free and frank atmosphere without any fear or inhibitions.

The interviews were conducted at places where respondents were available, which included their home, fields or their place of work. Information obtained was recorded immediately in the interview schedule.

Observation

At the time of investigation several visits were made from field to field in all the three villages. The study is based on one year's enquiry. It was conducted during the agricultural year 2003-04 expanding from 1st April, 2004 to 30th September, 2004, covering all the farm operations. Every possible care was taken to ensure the accuracy and reliability of the collected information. The informations furnished by respondents were properly edited through personal observation and cross-checks. The help of Chief Executive Officer, (Janpad, Mohkhed), Assistant Development Officer, Rural Agricultural Extension Officers, Village Sarpanch, (an Ex and recently elected). Old teacher and other village leaders were sought for obtaining correct and reliable data.

Group Interview

Different groups of progressive farmers, women and rural youth were formed in each sample village for group discussion in order to collect desired information and cross-check the collected data. The contract farmers, key informants and adopted farmers of KVK helped in the group discussion. In the course of group discussion, several meetings were held from time to time at different places in all the three villages. Maximum points were discussed with these groups. RAEOs, the contact farmers, key informants and adopted farmers of KVK, helped in the collection of data at the time of group discussions and also encouraged other farmers, landless and agricultural labourers to extend needful co-operation to the researcher.

Discussion with KVK Functionaries

The study is mainly based on primary data, but secondary data were also collected and used. Primary data was collected from the selected farmers/respondents. The same were verified from KVK scientists and extension personnel of KVK, Chhindwara. After collection of data from respondents some points were discussed with KVK scientists, regarding seed distributions, input supports, dissemination of new technologies, trainings, selection of farmers, selection of fields for demonstration, and farm demonstration. Demonstrations of different crops at the farmers' fields and extension activities through TOT programme were also discussed.

Secondary data were obtained from records of the KVK, Chhindwara. These reports included: Status Report of KVK (1983-84 to 2000-01); List of adopted villages; Photographs; List of conducted activities of sampled villages; and List of adopted farmers, contract farmers and key informants.

In short, villagers of all three villages under study are totally dependent on agriculture for their livelihood. They are from lower castes and lower and lower middle class. Relevant data were collected with the help of both primary and secondary sources. Interview schedule, observation and case study were the main techniques to collect data from primary source. Census report and published and unpublished reports/articles of KVK Jabalpur and Chhindwara were consulted to collect secondary information. Logical technique was used to examine and explain impact of KVK on farm work, income and social and cultural institutions and practices.

CHAPTER – FOUR

Respondents Profile

Chhindwara is one of the underdeveloped districts in M.P. in terms of availability of good quality land for agriculture operation and dependable source of livelihood, per capita income, rate of industrialization and urbanization, state provided infrastructural facilities and so on. The district is tribal dominated. Also, significant part of the geographical area is under forest. Since, people at large live in villages and their source of livelihood is mainly land, as a result of poor quality of land coupled with lack of irrigation and proper marketing facilities, significant number of population are below the poverty line. Needless to mention that about 84 per cent respondents in the study villages live in either *kachcha* or mixed type of houses. There are 754 households in all the three villages of Simariya, Jaam and Chargaon. Jaam is thickly populated in comparison to the remaining two villages. In all the three villages male outnumber the female population and children and young outnumber the adult and senior citizens. The number of widows and divorcees are significant in all the villages.

So far caste background is concerned, all the villages are numerically dominated by intermediate caste like Kunbi and Pawar. Tribal do not reside in these villages. This is clear from the following Table 4.1.

So far as occupation is concerned, farming and farm related activities are their main sources of livelihood. But wage and farm work is also done by about 14 per cent of population. The number of villagers either working in government or private Sector is

Table 4.1: Caste/Community-wise households

Sl. No.		Simariya		Jaam		Chargaon		GT	%
		Total	%	Total	%	Total	%		
1.	Kunbi	53	35.3	261	73.7	2	0.8	316	41.9
2.	Pawar	24	16.0	8	2.3	206	82.4	238	31.6
3.	Mehra	11	7.3	12	3.4	4	1.6	27	3.6
4.	Nai	9	6.0	6	1.7	0	0.0	15	2.0
5.	Mali	40	26.7	0	0.0	0	0.0	40	5.3
6.	Chamar	4	2.7	10	2.8	19	7.6	33	4.4
7.	Drum Beater	3	2.0	3	0.8	2	0.8	8	1.1
8.	Sonar	1	0.7	2	0.6	9	3.6	12	1.6
9.	Others (Muslim)	5	3.3	52	14.7	8	3.2	65	8.6
	Total	**150**	**100**	**354**	**100**	**250**	**100**	**754**	**100**

insignificant. More than 32 per cent households have either or more than the income of Rs. 11000 per year. In other words, there are 136 villagers out of 1700 in Jaam and 58 persons out of 1263 in Chargaon, who earn more than Rs. 50000 from agricultural work.

Availability of land and constant income is related to each other. The problem of landlessness is insignificant in all the three villages and about 75 per cent households have their own land, some have land on share/contract basis and only few are landless.

Table 4.2: Land Availability

Sl. No.	Land Size (in acre)	Simariya		Jaam		Chargaon		GT	%
		Total	%	Total	%	Total	%		
1.	Landless	23	15.3	46	13.0	44	17.6	113	15.0
2.	Up to 2.5	30	20.0	45	12.7	60	24.0	135	17.9
3.	2.5 to 5	36	24.0	36	10.2	16	6.4	88	11.7
4.	5 to 10	32	21.3	96	27.1	66	26.4	194	25.7
5.	10 to 15	18	12.0	44	12.4	37	14.8	99	13.1
6.	15 to 20	11	7.3	24	6.8	21	8.4	56	7.4
7.	20 to 25	0	0.0	30	8.5	6	2.4	36	4.8
8.	Above 25	0	0.0	30	8.5	0	0.0	30	4.0
9.	45 to 50	0	0.0	3	0.8	0	0.0	3	0.4
10.	**Total**	**150**	**100**	**354**	**100**	**250**	**100**	**754**	**100**

Irrigation is an essential input component for high yielding. As stated, there are 754 households in the villages, but 450 said that they have irrigation facility. They irrigate land primarily from well with the help of electric or diesel pump.

Table 4.3: Sources of irrigation (Multiple Response)

Sl. No.	*Land Size*	*Simariya*		*Jaam*		*Chargaon*		*GT*	*%*
		Total	*%*	*Total*	*%*	*Total*	*%*		
1.	Nothing	52	34.7	113	29.3	95	38.0	260	33.1
2.	Well	96	64.0	257	66.6	150	60.0	503	64.0
3.	Cannel	0	0.0	8	2.1	5	2.0	13	1.7
4.	Nalla	2	1.3	8	2.1	0	0.0	10	1.3
	Total	**150**	**100**	**386**	**100**	**250**	**100**	**786**	**100**

Villagers under study are neither illiterate nor highly educated. About 42 per cent villagers have received education up to high school. This does not mean that there is no expansion of higher education in the villages. About 7 per cent of the population have received education up to college/university level.

Cultivation

Barring landless or those who are wage earners, villagers do their farm work by themselves. Their number is 65 per cent in Simariya, 31 per cent in Jaam and 51 per cent in Chargaon. About 32 per cent do farm work with the help of labour and 6.8 per cent maily depend on labour for their farm work. Their number is relatively higher in Jaam in comparison to the remaining two villages.

Household/Agricultural Assets

So far as agricultural assets are concerned one can easily see availability of both traditional assets and modern assets in all the three villages in the proportionate manner. For instance, ox is available with more than 59 per cent households in all the three villages. Also, bullock cart is available with 45 per cent households, and cow, ox and buffalo are available with 43 per cent households. Well is even today the main source of irrigation in villages. It is available with 61 per cent households. But at the sametime pump set is also available with 41 per cent households in Chargaon and more than 60 per cent households in the remaining two villages.

Further, electricity and food storage facility is available with large number of respondents. Availability of Radio, T.V. and Cycle with large number of respondents shows their inclination in favour of consumer culture. But toilet facility is available with lowest 18.4 per cent households in Chargaon and highest 38 per cent in Jaam

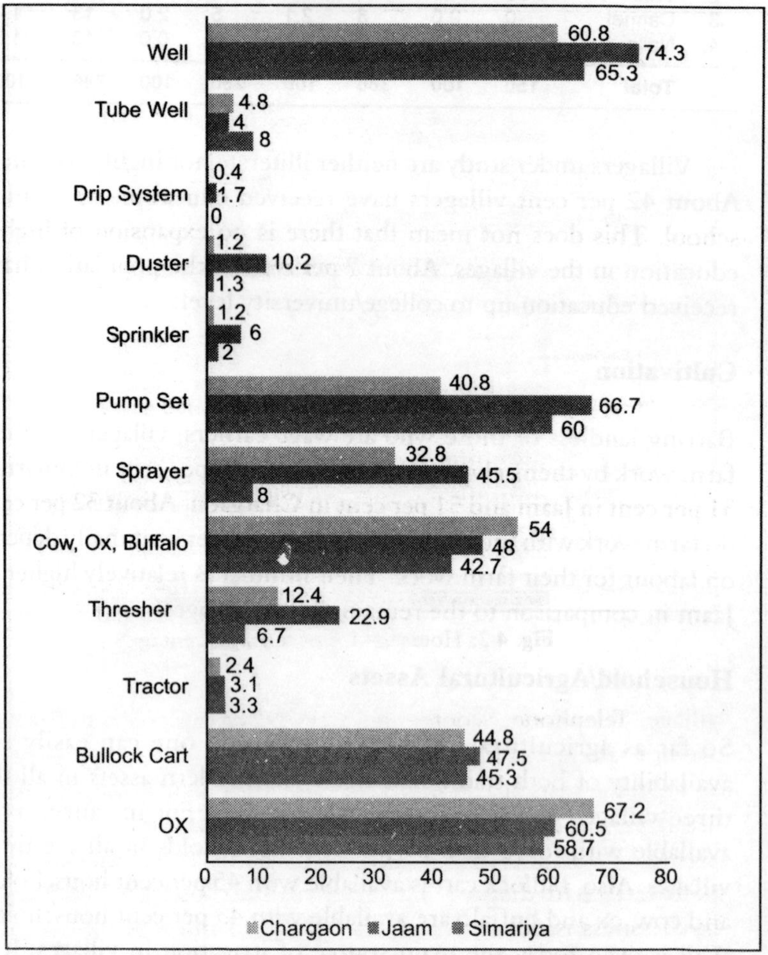

Fig. 4.1: Agricultural Assets (in percentage)

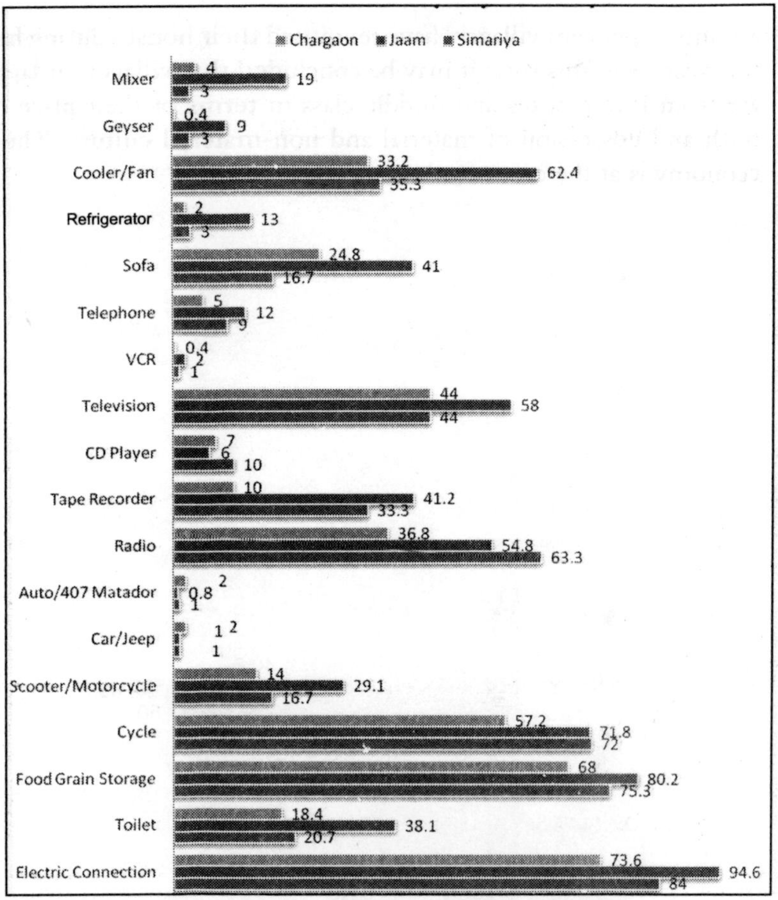

Fig. 4.2: Household Assets (in percentage)

village. Telephone, Scooter and Motor Cycle, Cooler, and Fan etc. are not rare in the villages.

On the basis of above analysis it is clear that majority of villagers are from the backward castes. The villages are underdeveloped in terms of education but young in terms of availability of young population in disproportionate manner and the problem of landlessness is not serious in the villages. Land is the main source of livelihood for the entire villagers. Needless to mention that around 77 per cent villagers have renovated their houses during last few years. On this basis it may be concluded that villagers at large are from lower castes and middle class in terms of their place of birth and possession of material and non-material culture. Their economy is at the cross-roads.

CHAPTER – FIVE

Exposure to and Adoption of KVK

Swaminathan (1999) making comments on expansion of KVK during last 20 years said that it is a technique which believes in learning by doing. This means technology introduced by KVK is made in laboratory. It is experiment based and soon after reaching to the farm it is gradually modified by the change agents and the farmers themselves in the light of its suitability and requirement.

In case the newly introduced technology is simple and friendly to the local environment and condition, there is a possibility of high level of knowledge or information among the users which subsequently make it favourable from the point of view of adoption and use of these techniques. Hence, it may be said that exposure is a necessary condition for useful and sustainable adoption of any technology for longer period.

The history of introduction of KVK is not very old in the villages under study but about 78 per cent villagers know about the existence and functioning of KVK in their village. Their number is highest in Jaam and lowest in Chargaon. In Chargaon those villagers who are living in the remote part of the village, significant number of them do not know anything about KVK. They are not in contact with KVK and therefore, the level of its adoption is very low among them.

So far as functions of KVK is concerned, respondents provided multiple responses. However, the number was much higher of those who said that it provides technical knowledge and detailed information about scientific farming. Some respondents also said that proper quality of seed is provided by KVK. The remaining respondents observed that it provides demonstration on farm, pesticide, training about seed and diseases, information about quality of soil, techniques to protect standing crops and so on.

Table 5.1: Functions of KVK (Multiple Response)

Sl. No.		Simariya		Jaam		Chargaon		GT	%
		Total	%	Total	%	Total	%		
1.	No Response/No Reply	13	6.3	44	9.9	122	44.0	179	19.3
2.	Information about farming	45	21.7	82	18.4	39	14.1	166	17.9
3.	Technical Knowledge	37	17.9	154	34.6	33	11.9	224	24.1
4.	Provide Seed	92	44.4	62	13.9	16	5.8	170	18.3
5.	Organise demonstration on farm	0	0.0	1	0.2	2	0.7	3	0.3
6.	Provide insecticide killers	0	0.0	7	1.6	14	5.1	21	2.3
7.	Training about seed/disease etc.	0	0.0	17	3.8	14	5.1	31	3.3
8.	Information about scientific farming	16	7.7	50	11.2	34	12.2	100	10.8
9.	Assess standing crop	1	0.5	1	0.2	1	0.4	3	0.3
10.	Seed demonstration	3	1.4	27	6.1	2	0.7	32	3.4
	Total	**207**	**100**	**445**	**100**	**277**	**100**	**929**	**100**

So far as villagers perception about functioning of KVK is concerned, 56 per cent replied in the affirmative. Their number was highest in Jaam and lowest in Chargaon. This shows that for majority of villagers KVK is functional and beneficial. Only 8 per cent respondents perceived its functioning in the negative way. Also, those who have not heard or have not availed its benefit, they could not pass comments on nature of its functioning.

Those who perceived its functioning positively gave number of examples to justify their point of view. For instance, they said that KVK provide new knowledge in time, it popularizes farming, poors are encouraged to seek benefit, encourage training, promote contact, KVK functionaries visit the village every now and then and it helps people to increase their level of technical knowledge. About 26 per cent respondents do not see any positive or negative impact. But very small number of respondents opined that its functioning is negative for some of the villagers. It was argued that it do not provide seed to everybody, particularly the poor and the needy, it do not work properly and it mainly extends benefit to only the rich and the favourite.

Table 5.2: Nature of Functioning of KVK (Multiple Response)

Sl. No.		Simariya		Jaam		Chargaon		GT	%
		Total	%	Total	%	Total	%		
1.	No Response/No Reply	9	5.6	71	19.0	132	50.6	212	26.7
2.	New knowledge in time	16	9.9	24	6.4	20	7.7	60	7.6
3.	Popularize farming	2	1.2	5	1.3	14	5.4	21	2.6
4.	Technical knowledge about farming	13	8.0	96	25.7	32	12.3	141	17.8
5.	Poors are denied benefits	18	11.1	15	4.0	14	5.4	47	5.9
6.	Everybody avail benefits	10	6.2	21	5.6	6	2.3	37	4.7
7.	Provide techniques of farming	5	3.1	18	4.8	3	1.1	26	3.3
8.	Provide training	1	0.6	22	5.9	1	0.4	24	3.0
9.	Establish Contact	0	0.0	5	1.3	4	1.5	9	1.1
10.	Help all	7	4.3	6	1.6	4	1.5	17	2.1
11.	They come most often	0	0.0	30	8.0	0	0.0	30	3.8
12.	Teach how to use seed/ fertilizer/pesticide	4	2.5	7	1.9	9	3.4	20	2.5
13.	Provide technical knowledge/instruments	4	2.5	11	2.9	2	0.8	17	2.1
14.	Inform about new variety of seed	3	1.9	1	0.3	3	1.1	7	0.9
15.	Level of technical knowledge have improved	0	0.0	1	0.3	1	0.4	2	0.3
16.	Do not provide seed to anybody	3	1.9	1	0.3	1	0.4	5	0.6
17.	Only favourites receive seed	6	3.7	0	0.0	0	0.0	6	0.8
18.	Needy do not receive seed	4	2.5	0	0.0	0	0.0	4	0.5
19.	Provide training in each month	5	3.1	6	1.6	3	1.1	14	1.8
20.	Do not work properly	0	0.0	1	0.3	2	0.8	3	0.4
21.	Provide seed only to rich	7	4.3	5	1.3	4	1.5	13	1.6
22.	Others	45	27.8	27	7.2	6	2.3	78	9.8
	Total	**162**	**100**	**373**	**100**	**261**	**100**	**793**	**100**

In other words, about 21 per cent respondents are of the opinion that both the rich and the poor do not receive benefit in an equal manner. Why it is so? Number of reasons were provided by villagers for such discrimination. For example, for many, lack of resources do not allow poor to adopt scientific farming. Lack of contact and time and interest constraint also discouraged many farmers to avail its benefits. Further, there are some lazy farmers or some have very small size of land holding, therefore, they are unable to avail benefit. Further, lack of irrigation, information, road, communication, etc. prohibit them to seek its benefit. Many persons have land in the interior area of the village where KVK functionaries do not visit and some of them have very poor quality land. As a result, they are unable to avail benefit of the project.

In spite of the small history of KVK in the area, significant respondents have participated in its activities. Till date about 40 per cent respondents have participated in its activities during last three years. Some of the important activities in which they have participated and availed benefit are related to vegetable production, peanuts and soya production, gram, masoor, *ratanjot*, urad and moong production, demonstration programme relating to crop treatment, poultry, and wheat production. Some of the farmers received the benefit of soil testing. Some attended workshop and visited *Kisan Mela*. Some also received fertilizer and learnt techniques to mitigate common problems relating to overall production of crops.

In response to a question that who encouraged them to avail benefit of KVK programme, the role of single agent like *Gram Sawak* or Sarpanch or progressive farmers or KVK scientists was not dominant. Less than 40 per cent farmers were encouraged by these institutions to avail benefits.

Those who did not avail benefit of KVK they said that due to lack of awareness, lack of resources including time constraint, poverty, and small land holding size etc., they could not avail benefits of KVK.

Training/Motivation

There is a positive interface between training, motivation and success of any development project. There is a point of view that even without questioning the historical and structural constraints such as lack of political will, manipulation, corruption, etc. at the programme implementation level just by imparting frequent training and motivating beneficiaries, the amount of success of the project can be enhanced. This hypothesis has been tested through number of studies on programmes relating to panchayat and other grassroots level institutions. For instance, Kumar and Singh (1994) said that the Government of India has increasingly emphasized upon training ever since the launching of the Community Development Programme. However, a giant level and direction with the initiation of KVK or farm science centre by the ICAR was hailed as an organizational innovation with its roots in time tested principles of training dovetailed with the best and all the essential components of non-formal education, viz. technically sound trainers, emphasis on skills development, well-endowed infrastructure and fund for training as also vertical and horizontal organization linkage.

In our study about 38 per cent respondents (i.e. 287) said that they have attended either workshop or training or both which was organized by KVK. About 275 respondents praised the programme and the remaining few were either critical or indifferent on the merits of such programme.

But about 60 per cent respondents gave suggestions to enhance effective functioning and success of training programme. They advised that if maximum information about new techniques is provided, practical demonstration is imparted, experience of big or successful farmers are shared by farmers themselves, training inside or at the village is organized, information is provided on constant basis, maximum villagers are encouraged to participate in training, training time is fixed as per the convenience of villagers and if there is effective involvement of panchayat, it will ensure success of training programme. Similarly, some of the respondents also said that if Kotwar is directed to inform villagers about training in time, training is imparted on the farms of all irrespective of their socio-economic background, sufficient and quality-ridden seed is frequently distributed in time, and farmers fair and exhibition is organized, it will enhance the level of awareness as well as success of training/demonstration programme.

Table 5.3: Suggestions to ensure success of training programmes (Multiple Response)

Sl. No.		*Simariya*		*Jaam*		*Chargaon*		*GT*	*%*
		Total	*%*	*Total*	*%*	*Total*	*%*		
1.	No response/No reply	19	11.7	162	42.1	112	44.1	293	36.6
2.	Provide maximum information about new techniques	7	4.3	16	4.2	5	2.0	28	3.5
3.	Practical demonstrations	1	0.6	20	5.2	3	1.2	24	3.0
4.	Information about new seed/fertilizer	2	1.2	5	1.3	4	1.6	11	1.4
5.	Experience of big farmers should be shared	1	0.6	10	2.6	0	0.0	11	1.4
6.	Farmer's capacity based information should be provided	2	1.2	7	1.8	0	0.0	9	1.1
7.	Training inside the village	8	4.9	29	7.5	11	4.3	48	6.0
8.	Information constantly	30	18.5	29	7.5	9	3.5	68	8.5
9.	Maximum participation in training	1	0.6	2	0.5	5	2.0	8	1.0
10.	Training/visit in time	0	0.0	10	2.6	5	2.0	15	1.9
11.	Panchayat's involvement	0	0.0	3	0.8	0	0.0	3	0.4
12.	Help of Panchayat should be sought to ensure presence	0	0.0	6	1.6	3	1.2	9	1.1
13.	Kotwal should inform about meeting date and time	0	0.0	1	0.3	0	0.0	1	0.1
14.	Training on farms of all	5	3.1	4	1.0	0	0.0	9	1.1
15.	Distribution of seed again and again	3	1.9	1	0.3	0	0.0	4	0.5
16.	Seed/fertilizer distribution free of cost	1	0.6	0	0.0	0	0.0	1	0.1
17.	Farmer's fair should be organized	2	1.2	2	0.5	1	0.4	5	0.6
18.	Demonstration should be regular	2	1.2	1	0.3	0	0.0	3	0.4
19.	Equitable distribution of seed	1	0.6	0	0.0	0	0.0	1	0.1
20.	Others	77	47.5	77	20.0	96	37.8	250	31.2
	Total	**162**	**100**	**385**	**100**	**254**	**100**	**801**	**100**

Farmers who have availed benefits of KVK also acknowledge the role of motivating factors in this context. About 23 and 18 per cent respondents praised the role of scientists and progressive farmers in this context respectively. Role of Sarpanch, *Gram Sewak*, Rural Agriculture Extension Officer (RAEO) etc. was also narrated by some of the villagers.

Here, it is needless to mention that long back Desai (1969) had said that Community Development Project (CDP) failed primarily because it relied more and more on the big farmers but they were selfish and they did not take interest in these issues. But in this study the finding is contrasting. This study admit positive role of both progressive farmers and scientists in the effective functioning of KVK. RAEOs of all the three villages expressed views in the following manner: "we are also attending the KVK seasonal training, local workshop with KVK scientists and prepare an action plan. According to this plan the technical programmes were held in these villages. Distribution of improved seed, improved implements and fertilizers was also made and farmers were encouraged to use these techniques. All these three villages are very progressive in our Block. Farmers are aware to get better production. KVK diffused technical information among farmers and we are also using these techniques. Our regular contact with farmers is intact. We have worked more for the improvement of these villages. KVK is also in the field but we also work with the help of the department schemes. In the programmes of KVK we are also involved. If there is any problem such as infestation of disease, severe attack of insect and deficiency of any micronutrients, then we always organize visit with KVK scientists. As a campaign we spread the solution of the problem as per the guidance of scientists. The guidance is used by the people of these three villages and get the gains by the solution because they have small holdings, agriculture is one and the only support base for them. They are always anxious to use new techniques."

Visit of KVK Scientists/Personnel

One of the objectives behind the establishment of KVK in the nearby area of the villages was to encourage frequent interaction between villagers and KVK scientists so that farmers can get maximum benefit of their guidance. One of the techniques to encourage interaction between the two is to make frequent visit to the villages by the KVK representatives, see the land management activities and provide suitable resources, including suggestions, which can functionally suit to the local conditions. Unfortunately, this possibility does not exist in the villages under study. Hardly 26 per cent respondents admitted that KVK scientists make frequent visit to their villages. Moreover, in Chargaon village only 9 per cent respondents made such comments.

Question was asked from these 26 per cent, i.e. 199 respondents about reasons behind their visit. Number of responses were recorded. It was replied that in order to watch standing crop, hear problems, provide treatment measures, suggest proper time for irrigation, provide technical knowledge, etc., visits were made. Some also said that visit was made to encourage farmers to produce green vegetables and garlic. This shows that of course their visit at least during last five years was not popular but it was frequent and beneficial for about 25 per cent of the total respondents. This is supplemented on the basis of following observations made by villagers.

Respondents said that had KVK representatives not frequently visited the village the farmers may not have received new information regarding modern farming. This visit and suggestion paved the way for modern and improved farming, solution to farm related problems, protection of crop and preservation of cereals and so on. Significant number of respondents learnt to produce specific variety of farming in the absence of proper irrigation facility..

As stated earlier, less than half of the respondents reported one or another type of demonstration done by KVK representatives on their land. The nature and type of demonstration was although different but every such demonstration was directed to encourage production, both for market and for self-consumption. For example, they heard from them about new variety of gram, soya, onion, tuar, moong and so on. Garlic demonstration and demonstration of groundnut was highly beneficial for some of the respondents. Some respondents learnt about production of bio-fertilizer and proper use of pesticides.

Table 5.4: Technological demonstration at farm during last 5 years (Multiple Response)

Sl. No.		*Simariya*		*Jaam*		*Chargaon*		*GT*	*%*
		Total	*%*	*Total*	*%*	*Total*	*%*		
1.	No response/No reply	28	14.6	66	16.4	178	69.0	272	31.9
2.	No demonstration	64	33.3	200	49.8	52	20.2	316	37.1
3.	Treatment of crop	0	0.0	2	0.5	7	2.7	9	1.1
4.	Information about new gram variety	24	12.5	10	2.5	4	1.6	38	4.5
5.	Information about new soya variety	22	11.5	55	13.7	1	0.4	78	9.2
6.	Demonstration relating to all crops	5	2.6	40	10.0	1	0.4	46	5.4
7.	Demonstration about *Fulgobhi* (cauliflower)	0	0.0	2	0.5	0	0.0	2	0.2
8.	Onion production	2	1.0	0	0.0	0	0.0	2	0.2
9.	Mustard production	5	2.6	2	0.5	0	0.0	7	0.8
10.	Moong demonstration	0	0.0	9	2.2	0	0.0	9	1.1
11.	Proper use of pesticides	3	1.6	2	0.5	2	0.8	7	0.8
12.	Wheat production	24	12.5	5	1.2	1	0.4	30	3.5
13.	Tuar production	0	0.0	1	0.2	3	1.2	4	0.5
14.	Varsim production	0	0.0	0	0.0	1	0.4	1	0.1
15.	Production of bio-fertilizer	2	1.0	1	0.2	1	0.4	4	0.5
16.	Achar production	3	1.6	0	0.0	0	0.0	3	0.4
17.	Garlic demonstration	2	1.0	0	0.0	0	0.0	2	0.2
18.	Onion demonstration	1	0.5	0	0.0	0	0.0	1	0.1
19.	Other (Crop demonstration)	5	2.6	2	0.5	3	1.2	10	1.2
20.	Groundnut production	2	1.0	5	1.2	4	1.6	11	1.3
	Total	**192**	**100**	**402**	**100**	**258**	**100**	**852**	**100**

Improvement in Farm Production

One of the immediate objectives of KVK introduction was purely economic in nature. It was supposed to enhance farm production and subsequently contribute to economic development at the family level. About half of the respondents observed that KVK intervention improved farm production. This became possible not just due to their suggestion but also because as per their advice land was improved and seed and fertilizer were applied. Also, KVK officials, frequent visit to farm land helped villagers to use technical knowledge in an assured manner.

But few respondents observed that KVK intervention was harmful to them. For instance, about 8 per cent said that due to crop failure they had to meet loss. Case of garlic loss was also reported by few farmers. Some villagers observed that lined farming reduced production. Some said that improved seed was provided late in the season which resulted decrease in farm produce. Santosh Gabde and Mahadeo Sadrang of Jaam village said that more infestation of disease, i.e., rust on soybean, gram, catter piller attack on gram and abnormal condition of weather resulted in crop failure. On this basis it seems that KVK was functional for many but it was also dysfunctional for at least some of the households.

Change in Cropping Pattern

About half of the respondents not only reported improvement in their farm produce after the introduction of KVK in their respective villages but many of them also reported change in the cropping pattern. Today production of garlic, vegetable, etc. has become very common particularly by the relatively big farmers. In the past they used to produce only wheat and soya but now it has been significantly replaced by the production of vegetables like ginger, tomato, coriander, green chilly, *bhindi*, cauliflower, onion and so on. In the rabi crop pea, potato and garlic is very common. Farmers produce these crops for market because all the three villages are duly connected to Nagpur market through a metalled road which passes through these villages. Farmers easily transport such farm products to the Nagpur market along with Chhindwara and Bhopal market regularly. This is also supported by the fact that in comparison to the past frequency of their visit to Nagpur and such other places have increased several times.

Along with change in cropping pattern today the area under crop have also increased. Villagers not only use their fallow land but there are some, mainly the rich, who have encroached community land for cultivation purpose. Rai (1995) observed that land utilization area under double crop has increased by 82.34 per cent and under triple crop area by crops after three years is 78.04

per cent. It has its positive impact on total farm produce and the economic condition of beneficiaries.

Panchayat/NGO and Functioning of KVK

Till date three panchayat elections have been held in M.P. under the 73rd Constitutional Amendment provisions. Thousands of women and men of Scheduled Castes and Scheduled Tribes have joined Panchayati Raj Institutions (PRI) at all the three levels in different capacity. Also, there is a saying in Madhya Pradesh these days that there is Non-Governmental Organisation (NGO) in the state. How these two interventions have affected success of KVK? In course of this study no significant role of these two institutions was reported by majority of respondents. Less than 7 per cent respondents narrated the effective role of panchayat in this context. But none of the respondents reported any role of NGO in their agricultural development. Some respondents also underlined role and importance of panchayat initiatives in the success of KVK.

Villagers Perception about Functioning of KVK

Respondents have expressed mixed opinion about overall functioning and impact of KVK technology. For example, about 75 per cent respondents said that KVK has not only popularized simple agricultural technology among cross-section of farmers but also it has solved many farm related problems which were otherwise unresolved in the past. Who can undermine importance of cash crop popularized by it in villages? But at the same time more than 50 per cent villagers are of the opinion that KVK is limited to only big farmers and even among them its style of demonstration is faulty. More than 60 per cent respondents agree with the fact that scientists constantly visit the villages and make sure that only appropriate technology is provided to the villagers. About 70 per cent respondents said that KVK has contributed to both agricultural modernization and economic development in their life. It has also popularized modern techniques of farming among about 71 per cent villagers. It was reported by about 44 per cent respondents that

Table 5.5: Panchayat/NGO and Functioning of KVK (Multiple Response)

Sl. No.		*Simariya*		*Jaam*		*Chargaon*		*GT*	*%*
		Total	*%*	*Total*	*%*	*Total*	*%*		
1.	No response/No reply	25	15.63	110	30.81	79	30.38	214	27.54
2.	Panchayat should organize exhibition/ workshop in collaboration with KVK	0	0.00	29	8.12	4	1.54	33	4.2
3.	Panchayat should prepare list of beneficiaries	0	0.00	10	2.80	10	3.85	20	2.57
4.	Panchayat already extend support	4	2.50	21	5.88	35	13.46	60	7.72
5.	Panchayat is redundant	0	0.00	7	2.0	2	0.8	9	1.2
6.	Provide place for organization of programme	8	5.00	3	0.84	1	0.38	12	1.54
7.	Construct small dam, provide facility for irrigation	2	1.3	2	0.6	4	1.54	8	1.0
8.	Provide raw material to built house on cheap cost	2	1.3	1	0.3	2	0.8	5	0.6
9.	Encourage farmers to participate in KVK fair/ Exihibition	1	0.63	12	3.4	2	0.8	15	1.93
10.	Only provide seed and no other role of panchayat in KVK work	4	2.50	5	1.40	1	0.38	10	1.3
11.	Not any help/not any role of KVK work	69	43.13	45	12.61	23	8.85	137	17.63
12.	Panchayat do not help	45	28.13	112	31.37	97	37.31	254	32.69
	Total	**160**	**100.00**	**357**	**100.00**	**260**	**100.00**	**777**	**100.00**

it has contributed to women's participation in farming, but 26 per cent had negative opinion on this issue. More or less similar number of respondents expressed mixed opinion about the usefulness and unusefulness of the related literature distributed among farmers by KVK representatives. More than half of the respondents said that not only they learnt about new techniques but they also shared such experience with co-villagers and relatives. About 20 per cent villagers admit that there was lack of understanding between the villagers and KVK representatives. To some respondents looking to the usefulness of KVK it can be said that it can also contribute to development in the future. People have faith in KVK representatives. About 62 per cent observed that scientists are not only highly informed persons, but they are also friendly and eager to share fruitful information with villagers. For them it is a successful programme although it has to become more and more poor friendly in the due course of time. In other words, about half of the respondents rated high opinion about its effective functioning. They also mentioned reasons of success or failure of KVK techniques. The factors which contributed to its success are, for example, use of new techniques, solution of problem, frequent demonstration, timely distribution of seeds, constant use of farm land without rest, etc.

At the same time it was also perceived in a negative way by some of the respondents. It was said that poors were not benefited, seed quality was not up to the mark, and those who lacked irrigation facility or owned small size of land, they could not get benefit and therefore, they highlighted negative impact. Govind Rao, a farmer from Simariya village, who holds about 5 acres of land said: *"Hamari kheti thori gaon ke ander hai, wohan vaiganik kabhi aye nahi, hamne kabhi beej nahi liya, prashikshan mila nahi, log kahte hain tumhara khet ander hai, hame koi labh nahi hua."* (My land is located inside the village, till date no KVK scientist have visited it. I never received seed. I never received training. Scientists hesitate to visit my land because it is inside the village, I did not receive any gain till date).

On the basis of above analysis it can be said that majority of villagers know about presence and functioning of KVK in their village. Majority of them have also availed its benefit and they have experienced the benefit of change in cropping pattern which has shifted from production for consumption to production for market. They have not only attended training programmes but also encouraged others to participate in the functioning of KVK. The role of KVK scientists, training programmes and frequent technical demonstrations by KVK representatives were perceived as enabling factors for the success of KVK project. However, it was perceived by a group of respondents that KVK have still to make its road among relatively poor farmers, who have small pieces of land and who reside in extreme remote part of the village. Also, KVK have still to reach to the landless and poor wage earners. And for this purpose it must introduce some poor and landless friendly schemes. However, success rate was significantly rated as high by most of the respondents.

CHAPTER – SIX

KVK, Agriculture Modernization and Income

With the introduction of KVK led technological modernization in agriculture the new or HYV seed, proper type of fertilizer and pests, scientific management of farm land, irrigation and patterns of cultivation became popular in all the three villages. However, in every aspect adoption and dissemination of new technology/ knowledge was not uniform either in all the three villages or among cross-sections of farmers of a singal village. The following analysis is an extension to this observation.

About 74 per cent respondents learnt and used improved variety of seed. It was used by maximum number of villagers in Simariya and Jaam but less than 50 per cent villagers from Chargaon could use KVK techniques. More or less similar trend exist in the case of use of fertilizer, pest, improved seed, agricultural instruments, etc. So far as removal of waste and unwanted grass, rocks, etc. from the farm land is concerned in this work also Chargaon lagged behind both the villages. So far as production of vegetable, fruits, and social forestry work is concerned in this case by and large uniform trend was observed in all the three villages. But popularization of proper technique of storage and marketing of vegetable and foodgrain was more scientific in Simariya in comparison to the remaining two villages. Water management was reported to be very effective in Simariya, but it was poorly managed in Chargaon.

About 72 per cent respondents with highest number in

Simariaya and lowest in Chargaon reported that because of adoption of new farm technology there is positive impact on farm production. Today people at large use fertilizer, new storage facility, new marketing system, new instruments, new water management techniques and so on. Use of these techniques and facilities were not very popular about fifteen to twenty years back when KVK was not in existence in the study villages. Also, knowledge relating to all these new interventions have become very popular among villagers irrespective of their caste and class background.

Income and Expenditure

KVK intervention increased farm production and influenced economic condition of villagers positively. For example, about 38 per cent respondents were under debt during the pre-KVK days. Their number was highest in Jaam and lowest in Simariya. The amount of loan taken by them varied from Rs. 500 to either or more than Rs. 20,000. About 21 per cent had taken loan between Rs. 5,000 and Rs. 10,000. The source of loan was more informal and less formal. But today persons under loan are less than 21 per cent. In the past only 49 per cent were not under debt, but today more than 65 per cent households are not under debt. Also, in the past person having loan of above Rs. 10,000 was more than 12 per cent but today less than 12 per cent have taken such high amount of loan. Today villagers prefer to take loan for production purpose. And because of such change their dependence on formal sources have increased and accordingly their dependence on informal sources like *mahajans* and *sahukars*, etc. has decreased. This shows that there is positive impact on their economic status because of the introduction of KVK technology, in terms of loan.

Saving

So far as status of saving during the pre-KVK days is concerned 69 per cent had no saving. Only 7 per cent had either or more than saving of Rs. 10,000. The source where they used to save was both informal and institutional. But today about 50 per cent households have saving. Hence, the number of households having no saving have decreased from 69 to 50 per cent. It is interesting to note that 8.5 per cent (64) respondents have saving of more than Rs. 30,000. Hence, not only the number of villagers having saving have increased over the time, but also number of persons with having huge amount of saving have also significantly increased. Also, in the past only 17 per cent households used to save their money at formal institutions but now their number have increased up to 33 per cent. This shows favourable inclination towards bank or post office and other institutional bodies.

Economic Condition

The above finding is also supported from the fact that in the past 5.4 per cent respondents perceived that their economic condition was good. Majority perceived that it was either bad or neither good nor bad. But today 45.8 per cent feel that their economic condition is improved. In other words, during pre-KVK days only 41 respondents perceived their economic condition in an excellent manner but today 345 households perceived their economic condition in that manner. This shows improvement in their economic condition. And for this they attribute to only KVK intervention.[1]

Many scholars have discussed the role of agricultural modernization in economic condition of the concerned households. For instance, Acharya (1973) said that technological change was considered as a change from traditional farming to production syndrome, encompassing use of HYV seeds, chemical fertilizer, insecticides and use of modern irrigation equipments and machineries like tractor and farm implements. He concluded that net farm income increased by 46.3 per cent from pre-technological changes situation to post-technological change situation. Similarly, Khan and Chandragowada (1999) in course of their study in Karnataka village observed that in one of the schemes 5 farmers were not only imparted training on vegetable production but also leaflets pertaining to cultivation was distributed among them. These programmes facilitated farmers to adopt recent improved technology. Significant among them belonged to the use of improved crop varieties. Use of balanced fertilizer and adoption of need-based plant protection facilities improved farm production which ultimately increased their farm based income.

However, the number of persons who have experienced improvement varied from village to village.

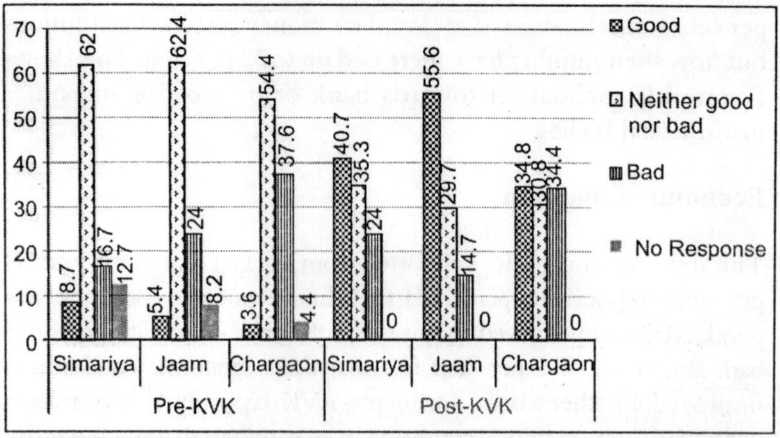

Fig. 6.1: Economic condition during pre- and post- KVK intervention

In response to a question that during last three years what were the factors which contributed to improvement in their economy, respondents narrated the role and importance of HYV seed again and again, use of scientific technique in farming, hard labour in farming, improvement in farm production, new information provided by KVK and so on. In other words, for majority of farmers (575) economic condition improved primarily because of KVK intervention, but in certain cases role of other factors also cannot be underestimated.

Those who explained the role of other factors during last three years attributed to the role of government job and such other external factors in the improvement of their income. But at the same time some of the respondents also said that because of lack of profitable size of land, lack of implementation of Minimum Wage Act and replacement of cash crop by crop for consumption, their income have decreased during last several years. Most of these respondents are landless wage earners.

Table 6.2: Reason of improvement in income (Multiple Response)

Sl. No.		Simariya		Jaam		Chargaon		GT	%
		Total	%	Total	%	Total	%		
1.	Landless/no gain/no reply	57	36.5	117	32.6	128	51.0	302	39.4
2.	Farm production increased	14	9.0	95	26.5	30	12.0	139	18.1
3.	Farm produce now sold on market rate	1	0.6	5	1.4	2	0.8	8	1.0
4.	Income increased due to increase in farm products	13	8.3	31	8.6	12	4.8	56	7.3
5.	Use of pesticide increased farm products	4	2.6	28	7.8	2	0.8	34	4.4
6.	Use of new technology increased farm products	42	26.9	50	13.9	47	18.7	139	18.1
7.	Favourable technique increased farm products	1	0.6	20	5.6	7	2.8	28	3.7
8.	Use of scientific technique increased farm products	3	1.9	7	1.9	12	4.8	22	2.9
9.	Information received and used at field	14	9.0	1	0.3	9	3.6	24	3.1
10.	Yield increased due to good quality seed	1	0.6	2	0.6	2	0.8	5	0.7
11.	Land given on rental basis	6	3.8	3	0.8	0	0.0	9	1.2
	Total	**156**	**100**	**359**	**100**	**251**	**100**	**766**	**100**

About 785 respondents i.e. maximum from Simariya and minimum from Chargaon admitted substantial improvement in their economic condition during last few years. More than 825 also admitted that KVK has not only improved economic condition of the individual households, but also it has improved village economy. This statement was made by 93 per cent respondents from Simariya and Jaam and 60 per cent respondents from Chargaon village[2].

Wage/Farm Employment

There are 15 per cent households in all the three villages who are landless. Majority of them are wage earners. Depending on requirement and conditions they work within and outside the villages. There is seasonal variation but in about half of the cases they are paid as per the rule. But women and children are less paid in comparison to the male workers. However, with an increase in farm income coupled with increasing cases of migration, they are paid more in comparison to the past.

More or less similar observation was made by other researchers in other parts of the country. For example, Raju (1982) using the regression on employment said that new agricultural technology had significantly contributed to an increase in the employment of labour. Among the different items of new technology, the application of fertilizer has significantly contributed to an increase in human resources.

On the basis of above analysis it is clear that KVK reached in all the three villages in different forms but it succeeded in the diffusion of new farm technology among majority of the villagers. As a result of this, it has increased their income and saving and has decreased the number of persons who were under debt. Further, their interest in institutional saving have also increased. Also, KVK have positively impacted over all economic condition of all the three villages both at the household and at the village level.

NOTES

1. Case study of Ramaraoji, Tonpe of Varjali village falling under Pandhurna block of Chhindwara district demonstrate that how a medium farmer improved his income by adopting cash crops under the influence of KVK Chhindwara. The story of Ramaraoji is basically a story of orange grower and has orange orchard. He has six acres undulated gravely soil land with no humus, no clay and no depth but he wanted to do something new. He visited KVK Chhindwara and approached KVK scientists and attended trainings and *Mela*. In 2009 KVK scientists advised him to grow Papaya in that land. He was convinced and started cultivating commercially viable Taiwan-786 Papaya in six acre land with drip irrigation. He also learnt about integrated nutrient management, integrated pest management and water management through drip irrigation. Today there are 4800 plants in his one year two months old Papaya orchard. He is getting 50 kg to 1 quintal Papaya per plant and he earns Rs. 25 lakh net and still he is taking produce. He is harvesting daily 4 tonne fruit of 1-4 kg

size and sells it in the market of Nagpur, Bhopal, Jammu and Chandigarh with Rs. 5-8 / kg. He started harvesting from December 2009 and now he is planning to increase the area up to 7 acre. By doing this he has generated employment of 10-12 labourers on regular basis. Also, he has encouraged 12-15 farmers to cultivate papaya in 22 hectare area. (*Source*: Obtained from http://www.kvkchhindwarazpdvili.org/success.php dated15/04/2013)

2. Similar observation was made by the author (Chaudhary, 2012) in other field based study in the nearby Betul district. It was found that over the time not only there was an improvement in the case of income of many households but also it was observed that there was an improvement at the village level.

CHAPTER – SEVEN

KVK and Social Change

There is difference of opinion among researchers on the issue of relationship between economic development or improvement in income both at the household and at village level and social change. If we have example of positive relationship between the two there are also examples of negative relationship between the two. In other words, economic development can lead to social change but at the same time it can also promote traditionalism. It can lead to institutionalization and effective functioning of democratic process but at the same time it can also frustrate democratic process and can facilitate primordialism. In a caste ridden society like India technology induced economic development can only lead to change in the value system and it may not lead to structural change. It may work for an increase in the age at marriage but it may not lead to eradication of dowry system. Rather in many cases economic development may contribute to *dowry* system. It may directly or indirectly contribute to expansion of education and occupational mobility but it may not lead to eradication of untouchability from the private sphere. Hence, like other issue technology led economic development have also two sides.

In this study, as discussed in the previous chapters, it has been analyzed that how KVK-led economic development have contributed to social change? How it is unable to promote social change? and, How it has inhibited social change? Here social change refers to both change in social structure as well as change in the value system of the structure. To me structural change in tradition-ridden villages is not possible. But change in the value system is possible and it is a common feature.

People's Perception

Majority of villagers are of the opinion that KVK has brought general improvement at the village level. The level of economic improvement, of course, have taken place at the individual level,[1] but such trend is more obvious and effective in Simariya and least beneficial and therefore, obvious in Chargaon village. It has also contributed to improvement in irrigation facility, improvement in and popularisation of technical knowledge, modern means of agricultural operation and living standard. They have moved towards modern assets. Further, it has changed food habit, dress pattern and capacity to purchase agricultural instruments. It has also improved education among children. Anand Rao, a teacher belonging to Simariya village said:

> "Today farmers consume more milk, green vegetables as well fruit, and *kheer*, *puri*, *halwa* and sweets which were prepared at the time of only festivals in the past. Now it is very common in day-to-day meal. Now-a-days, birthday, *mundan*, *katha*, marriage anniversary and small festivals are also celebrated with great joy and they prepare above items in each feast."

It is also interesting to note that because of modernization in agricultural activity and assured income from farm even among young generation attraction towards farming have substantially increased. Some of the respondents were of the view that even after receiving education they are unable to get job in either public or private sector. Therefore, they have concentrated in farm activities. Jamkar Lokhande of Jaam is employed in Land Development Bank. He said, "As improvement in agricultural production due to KVK intervention stared, I am giving more and more attention to farm activities."

More or less similar observation was made by Ram Khilwari of

Simariya village. It is needless to mention that KVK has substantially decreased seasonal outmigration. This is a good sign of social change. About 84 per cent respondents also reported improvement in their self-confidence due to interaction with KVK functionaries. Also, now they experience improvement in transportation facility even at the family level.

This does not mean that such development has taken place at the cost of social harmony and brotherhood in the village. For example, although 61.5 per cent said that amount of competition among villagers have increased over the time but this does not mean that such trend have negative impact. In response to a question that do they feel that today village is more disturbed and degree of rivalry have increased, 28 per cent replied in the affirmative and more than 65 per cent in the negative. Further, about 62 per cent disagreed that KVK-led development has contributed to corruption in the village. It is also obvious that health consciousness have increased and women's participation in farm activities have also substantially increased. Women not only participate in the production and gradation of cash crops but also their advice is respected in its marketing. Their requirements are met by obtaining required items when male members visit Nagpur and Bhopal for marketing of cash crops.

But negative consequences of development process were also noticed. For instance, about 74 per cent said that income disparity between big landowners and marginal farmers and landless workers have increased. This finding is similar to other studies which concludes that with the introduction of new technology or farm based development schemes, disparity between rich and poor or landowners and landless increases. The findings also reveals that heavy concentration on land and more and more income from land have increased land cost more in the case of land for house construction and less land cost for farm work. Further, amount of dowry have increased in all the three villages. It is highest in Jaam where KVK is more successful and lowest in Chargaon where it is less successful. Respondents also feel that quality of local leadership have improved and in comparison to the past today villagers have more and more access to modern means of communication.

Table 7.1: Social Change

Sl. No.		*Simariya*	*Jaam*	*Chargaon*	*G.T.*	*%*
1.	Improvement in living standard	144	325	183	652	4.6
2.	Improvement in food quality	146	320	178	644	4.5
3.	Improvement in education of children	146	327	188	661	4.7
4.	Interest in cultivation	145	311	188	644	4.5
5.	Self-confidence improved	145	307	182	634	4.5
6.	Competitiveness increased	129	216	119	464	3.3
7.	Village now disturbed	111	30	71	212	1.5
8.	Rivalry increased	114	30	67	211	1.5
9.	Corruption increased	122	46	84	252	1.8
10.	Health consciousness increased	143	287	156	586	4.1
11.	Women's participation in farm increased	147	301	182	630	4.4
12.	Dowry amount increased	133	338	180	651	4.6
13.	Leadership quality improved	140	304	174	618	4.3
14.	Now people send their children in town for education	149	329	175	653	4.6
15.	Social status improved	146	335	184	665	4.7
16.	Age at marriage increased	149	351	192	692	4.9
17.	Religious activities increased	148	342	177	667	4.7
18.	Change in sleeping and rising time	145	327	179	651	4.6
19.	Guest entertainment increased	148	334	198	680	4.8
20.	Women's self-confidence increased	145	331	185	661	4.7
21.	Small family norm more popular	136	345	185	666	4.7
22.	Untouchability decreased	129	346	183	658	4.6
23.	Gender disparity decreased	147	324	184	655	4.6
24.	Youth's migration decreased	146	309	192	647	4.6
25.	Crime against women increased	5	3	20	28	0.2
26.	Crime rate increased	4	4	6	14	0.1
27.	Cases of theft/robbery increased	2	3	4	9	0.1
28.	Police intervention increased	2	3	5	10	0.1
	Total	**3366**	**6828**	**4021**	**14215**	**100**

It is also clear from Table 7.1 that as per the perception of villagers there is no increase in the rate of crime against women, overall crime rate including theft and robbery and police intervention in villages. This is a healthy and positive sign of development. It is also interesting to note that in all the three villages interest in favour of children's education have increased and as a result, they send their children to even outside the village. It is needless to mention here that about 16 per cent members of respondents households are even today illiterate but 84 per cent respondents feel that KVK have improved education status of villagers. 92.3 per cent feel that relevance of education is highly beneficial in contemporary society.

This is significant change in people's perception and attitude towards education. Today majority of respondents (58%) wish to impart education to their children to the level as desired by children themselves. Even some respondents feel that the level and quality of education which they wish to impart should be high and quality ridden so that they may even become IAS and Engineer.

Table 7.2: Expected level of education to children (Multiple Response)

Sl. No.		Simariya		Jaam		Chargaon		GT	%
		Total	%	Total	%	Total	%		
1.	No response/ No reply	0	0	8	2.2	12	4.5	20	2.5
2.	As per their desire	135	84.4	243	66.2	74	27.5	452	56.8
3.	Graduation	0	0	10	2.7	10	3.7	20	2.5
4.	Technical course	0	0	4	1.1	0	0.0	4	0.5
5.	No gain/No teaching	2	1.3	5	1.4	1	0.4	8	1.0
6.	Up to 10	10	6.3	11	3.0	47	17.5	68	8.5
7.	Up to 12	6	3.8	37	10.1	54	20.1	97	12.2
8.	Up to M.A	2	1.3	9	2.5	28	10.4	39	4.9
9.	BE/IAS	1	0.6	2	0.5	2	0.7	5	0.6
10.	Job oriented course	1	0.6	10	2.7	12	4.5	23	2.9
11.	Others	2	1.3	19	5.2	15	5.6	36	4.5
12.	Childless family (Not applicable)	1	0.6	9	2.5	14	5.2	24	3.0
	Total	**160**	**100**	**367**	**100**	**269**	**100**	**796**	**100**

Their attitude in favour of girls education have also changed. They see number of benefits out of girls education. For instance, 17,16 and 13 per cent villagers said that it will help them in proper management of household activities, will increase the current level of commonsense and will help them to become self-dependant respectively. Significant number of respondents said that girls education is must because it will ensure their bright future, contribute to overall development of family and will also help them to avail job opportunity.

Table 7.3: Expected benefits of girls education (Multiple Response)

Sl. No.		Simariya		Jaam		Chargaon		GT	%
		Total	%	Total	%	Total	%		
1.	No response/No reply	3	1.8	8	2.2	31	11.7	42	5.3
2.	Proper management of household activities	13	7.6	100	27.8	26	9.8	139	17.4
3.	Job	18	10.5	20	5.6	32	12.0	70	8.8
4.	Self-dependant	20	11.7	43	11.9	40	15.0	103	12.9
5.	Marriage in respectful family	8	4.7	19	5.3	36	13.5	63	7.9
6.	Overall development of family	8	4.7	62	17.2	10	3.8	80	10.0
7.	They have right to be taught	21	12.3	5	1.4	6	2.3	32	4.0
8.	Level of knowledge will enhance	5	2.9	5	1.4	13	4.9	23	2.9
9.	Bright future	0	0.0	30	8.3	47	17.7	77	9.7
10.	Good housewife	9	5.3	6	1.7	1	0.4	16	2.0
11.	Development of commonsense	50	29.2	61	16.9	20	7.5	131	16.4
12.	Good home work	3	1.8	0	0.0	3	1.1	6	0.8
13.	Up to 12, so that some job may be obtained	13	7.6	1	0.3	1	0.4	15	1.9
	Total	**171**	**100**	**360**	**100**	**266**	**100**	**797**	**100**

Further, in comparison to the past today the use of HYV seed and fertilizer, pesticides, and use of material culture, etc., have significantly increased. But social institutions have also been positively influenced. 92 per cent respondents feel that age at marriage for both boys and girls have increased and there is no instance of child marriage in last couple of years in all the three villages. Their involvement in religious activity have also increased. Here, it is needless to mention that in another study in Sehore district of M.P. which is relatively close to the state capital, it was observed that villagers at large, irrespective of their caste background, are taking serious interest in temple construction. And this interest in temple construction has taken place during last several years. One of the possible reasons behind this may be improvement in their economic condition along with increasing amount of frustration and feeling of insecurity which have forced them to involve more and more in religious activities (Chaudhary, 2003).

It is observed that sleeping and rising time have also been influenced by agricultural modernization. They go on bed late in the night but they rise early to take care of domesticated animals. They are so enthusiastic about their income that they feel that if they will invest more, in due course farm income will increase. Today number of visitors and guests have also increased. It is also observed that small family norm has become popular and respondents at large feel improvement in confidence level of women. This is supplemented by the fact that according to 87 per cent respondents gender discrimination have decreased in the villages. In all the three villages degree of untouchability have also decreased in both public and private life. However, on certain occasions untouchability is still practiced. For instance, Jamkar Lokhande, Mudhukar and Bhaman Rao said that since it is a part of tradition, therefore, on specific occasions it is practiced by all. They said that *dalits do not sit* with the higher castes. One SC respondent said,

> "We are habitual of this. Late Tandekar (IAS) was extended respect by all villagers without caste consideration. This is because he was highly educated and betterly placed. But we do not, because we are poor and illiterate."

Eighty-three per cent reported that wage rate have increased but most of the wage earners observed that they are not paid as per the rule, although many of them do not know about the rule. Some of them are of the opinion that KVK intervention have improved economic condition of farmers, but they are not in favour of sharing this benefit with the wage earners, who are attached to them since long.

Respondents also identified some negative consequences which have taken place in all the three villages since last several years. However, they do not blame only KVK intervention for this state of affair. For instance, about 31 per cent said that pollution has increased, although 41 per cent differed on it, 93 per cent said that heat level has increased and number of them with no villagewise variation said that both surface water and underground water table have gone down. About 95 per cent said that during last several years number of plants and trees have decreased in villages. They also said that number of mosquitoes have rapidly increased. Respondents expressed mixed opinion about increase of milch cattle in the village. However, fortunately it was reported by only very small number of respondents that during last five years number of crimes in general including theft and robbery and crime against women in particular as well as police intervention in village matters, etc. have not increased. In short, whatever development is taking place in villages, these are, of course, mixed in nature but it is more functional and less dysfunctional.

Impact on Health and Disease

In order to deal with various diseases villagers simultaneously depend on different techniques since long. These techniques are both traditional and modern and both set of techniques are going on hand in hand. There is no conflict between the two. During the pre-KVK days 93.75 per cent respondents depended on local doctors. Most of these doctors were untrained and *Jhola Chhap*, today also 91 per cent of them mainly depend on them. Why most of them depend on them? In course of discussion with villagers it was reported that villagers have decades old trust on them. There is similarity in terms of social and cultural background among them. Also, there is similarity between patient and doctor in terms of their language and dress. Again easy availability of doctor, treatment on credit and sometimes free of cost, lack of modern treatment facility at the village level etc: motivate patient to avail service of *Jhola Chhap* doctor. *Jhar-Phuk* technique is also followed in all the three villages. However, number of its users have decreased from 14 households to 6 households. Similarly, cases and frequency of their visits to modern doctors at distant places have increased. For instance, in the past only 33 villagers visited doctors at a distant place during one year but today large number of them visit Chhindwara and Nagpur as and when medical help is required.

Table 7.4: Type of treatment availed

Sl. No.	Type/Place of Treatment	Simariya		Jaam		Chargaon	
		Then	Now	Then	Now	Then	Now
1.	No response	—	1	2	2	—	—
2.	*Jhar Phuk*	—	—	1	—	25	8
3.	Local untrained doctor	148	144	348	344	247	246
4.	External trained doctor (Chhindwara/ Nagpur)	10	15	15	13	8	24
	Total	**158**	**160**	**366**	**389**	**280**	**278**

It is needless to mention that with the increased frequency of their visit to Chhindwara and Nagpur to sell farm produce during the KVK period their attitude in favour of modern doctors and health services available in urban areas have positively changed. They have developed faith in it. Hence, it may be said that as their interaction with urban society will increase gradually their dependence on modern health services will also increase. Because today small number of villagers visit towns for commercial purpose, therefore, majority of villagers do not have exposure to modern health facilities.

On the basis of above observation it is clear that KVK not only increased income of majority of households but with the help of other interventions it has also brought about change in their physical quality of life. It has changed their attitude against out-migration. Due to frequent visit to towns and cities in order to sell their commercial farm produce the level of contact with outsiders as well as urban culture have increased which in turn have influenced their attitude towards health, education, dress pattern, means of entertainment, girls education, environmental cleanliness and so on. Attitude have also changed in favour of children's education, health and so on. It has discouraged crime including untouchability but it has contributed to dowry system. It has increased age at marriage, but it has also promoted economic disparity between different social groups without assuring wage payment as per the rule. Hence, it can be said that from social change point of view impact of KVK is mixed in nature.

NOTE

1. KVK is not restricted to only agricultural modernization. It is also engaged in the economic development of people by capacitating them in other walks of life. Following case of a woman may be taken as an example, Kabira Pawar (Divorcee) from Chandangaon village of Chhindwara district is a housewife. She is the only earning hand in the family. She is living with her father. Previously her only source of income was father's pension. Hence, she was forced to experience financial constraints. Anyhow she came in contact with KVK. And as per their advice she attended training on sewing at Chhindawara in 2007. After attending training she started her own boutique under the guidance of KVK scientists. Today by stitching clothes she is earning between Rs. 6-7 thousand per month. She is also imparting training of tailoring to rural girls and women. She has become the role model for rural women. (*Source*: Obtained from http://www.kvkchhindwarazpdvili.org/success.phpdated15/04/2013)

CHAPTER – EIGHT

Case Study

Discussion in preceding chapters clearly reflect mixed impact of KVK exposure on the level of knowledge of the concerned farmers, adoption and use of new technology and technical knowledge, farm production and subsequently income. Also, relationship between income, both at the individual and at the village level, and social change was discussed. It was also observed that many households who are poor and who live in the remote part of the village, they are mostly failed to avail benefits of KVK. Present chapter, Case Study, also support the findings discussed earlier. Although the cases described below are not representatives of the universe but in the form of qualitative narratives it enables us to the deeper understanding of relationship between new introduced farm technology, economic development and social change. Most of these cases, however, deals with success story.

Rameshwar Aldak

Rameshwar, a 46 years old and hard working person is an ambitious man. He belonged to joint family. KVK intervention brought affluency in the family in such a way that division of property took place in the family and now he hails from nuclear family. In course of division of property relatively poor and light soil ridden land was provided to him, but since he was a very labourious man, by dint of hard labour and help from KVK he could improve farm production and subsequently income. He experimented all modern varieties of crops. He attributes KVK exposure for change in cropping pattern on his farm in favour of cash crop like *bhindi*, carrot and garlic. He perceives himself fortunate because KVK intervention brought economic development in his life which enabled him (a) to get the post of Sarpanch in Gram Panchayat which he is holding, and (b) in the presence of about 500 farmers he was felicitated by KVK on the eve of birth anniversary of former P.M. Shri Charan Singh on December 23, 2002 for his best performance. The programme was organized at Simariya village itself.

The success story did not end here. Rather he was also identified as contact farmer for the villagers. He was supposed to promote and diffuse new farm technology, including knowledge, among co-farmers. Also, on the advice of KVK he formed Farmers Club. He prepared an exhaustive list of villagers land based demand and suggestions. These demands and suggestions were related to sowing method, spacing, time of sowing, seed treatment, use of proper quantity of fertilizer, use of micronutrients and use of insecticides. Farmers also expressed interest to know about use of fungicides for seed treatment, spray of insecticides, and maintenance of plant population.

Because of his intervention, KVK scientists on the one hand trained farmers on the above issues and on the other hand conducted demonstration on farm itself. The overall impact of this process was improvement in income of large number of villagers. He narrated that by seeing the benefit of its adoption by him other fellow farmers also got inspiration and adopted new technology. To him KVK conducted number of trainings, demonstrations and meetings at the village and farm level. He praised skills of scientists and their ways of communication. KVK not only organized meetings and training but also organized *Kisan Mela*, exhibition and *Kisan Sangoshti*. He also availed a number of study tours in Punjab, U.P., Bhopal and Jabalpur. Number of times he has participated in district level "Farmers Forum" activities.

Due to improvement in farm income he has constructed a *pacca* house and also one modern farm house. Now he owns TV, Sofa Set, Cooler and Refrigerator. He said:

> "I wish to become a social reformer and help farmers in every sphere to my best knowledge acquired by the KVK instructions."

Finally, he expressed happiness on the issue of ICAR award to the local KVK for its outstanding contributions during 2000-01.

Moreshwar Ghatakarde

Ghatakarde from Jaam village is in contact with KVK since last 7-8 years. He has 9 acres of irrigated land. He is satisfied with his annual income. He was impressed with KVK in such a way that he formed "Farmers Club" in the village. Regular visit to KVK office, attending training programme organized by them several times and observing demonstration activities, etc., have helped him to avail benefits of KVK as much as it is possible.

He stated that due to KVK intervention there is improvement in his economic condition. The condition of crop have also improved because of adoption of new technology. There is significant difference today in terms of use of HYV seed, use of fertilizer, micronutrients, insecticides for control of pest and diseases. He narrated that actually adoption of scientific technology by rich farmers encouraged him to adopt KVK technology.

Due to improved economic condition he repaired the old well, constructed a house and purchased a piece of land. In the past he and his uncle used to lift water from Dam by underground pipeline irrigation, which increased productions. But conflict emerged among them and now he irrigates land by well which he has repaired.

KVK scientists conducted field demonstrations on variety of crops like soybean, wheat and gram on the fields of rich farmers, particularly those firms which are situated on the road side. First he objected to this practice but he was convinced by the KVK functionaries that subsequently they will also work on other farms. Regarding the overall functioning of KVK he said that there was less coordination between agriculture department and KVK. Farmers of Jaam preferred and adopted the guidelines suggested by scientists whereas suggestions of agriculture department were preferred at the second level. However, there is substantial improvement in farm related scientific knowledge of farmers. He praised KVK for dissemination of agricultural technology through TOT (Transfer of Technology) programme in the district. KVK also distributed literature relating to applied technology of cultivation among farmers. He also remarked that rural youth have generated income through adoption of nursery technique for vegetable production in Jaam village. He contended that due to increased income he has deposited about Rs. 20,000 against the repayment of outstanding loan and got sanctioned Rs. 40,000 as a new loan for purchasing bullock cart and other agricultural instruments. Further, he has saved Rs. 60,000 in the local cooperative bank. Also, it was because of increased farm based income that he purchased a T.V. set, Motorcycle, Sofa-set, Tap recorder, Telephone and constructed modern toilet in his house. Also, he obtained a pump-set, sprayer, duster and thresher in the recent past. He strongly praised KVK which brought prosperity in his family life. Finally, he suggested that KVK should form small group of rural youth, women and small farmers to ensure success of its agenda on regular basis.

Anil Mankar

Anil is a young man of Kunbi caste who hails from Simariya village. He is a post-graduate degree holder youth. He attempted for government job but could not succeed. Finally, in consultation with KVK representatives he decided to engage in farm activities on permanent basis. In June 1997, he attended 3 days training programme on Kharif production technique which encouraged him significantly. Even today he actively participates in field visit, field demonstration, farmers interaction, selection of farmers and selection of demonstration activities. Being impressed by his interest and performance he was appointed as "Key Informant" by KVK. From KVK he learnt about Integrated Pest Management Technique (IPMT) which is applied on gram under the scheme called Trichoderma Viridi for seed treatment, neem oil, mpv, pheromone and insecticides (endosulfan) for control of Gram pod borer. Adoption of this technique increased farm production. It had demonstration effect on co-farmers. However, those who could not follow it became enemy of Anil.

Anil formed Farmers Club consisting of co-villagers who belonged to the same age group. Because of KVK teaching about 25 vermi-compost and the same number of nadep compost were constructed in Simariya village itself. Farmers of this village jointly organized a farmers fair and exhibition in December 2002. During this occasion KVK demonstrated different technologies like technique to identify insect and its impact, disease and weeds. It was a very effective programme from farmers points of view.

About the role of KVK representatives he said that they make frequent visit, collect facts and demonstrate need based techniques like linseed production technology, *kharif* onion production technology, techniques related to use of micro-nutrients in maize, soybean, cauliflowers, groundnut and spray of fungicide, seed treatment, methods of sowing and IPM in gram. As a result, demonstration area of land under the production of soya, gram and vegetables substantially increased in the village. He owns about 13 acres of land with one tubewell and one well for irrigation. Due to increased income he managed tubewell. Today the entire land is under irrigation. He grows vegetables like bringal, onion, coriander and chilly in a substantial manner. To him KVK widened economic gap between rich and the poor in the village. The degree of competition between farmers have increased. People have negative views about Anil. Anil said about perception of villagers about him in this way:

> "Only Anil of Simariya became the key informant and was given the opportunity to demonstrate prescriptions of KVK in the field."

But this allegation was refuted by member of farmers club like Arjun Charpe, Arun, Anand Rao and Niranjan. They said:

> "We formed a farmers club and participated in KVK activities like training, field demonstrations, field visit, field day, farmers fairs and farmers days, they were benefited under KVK programme."

Anil said that he is doing best to ensure participation of all co-villagers in the functioning of KVK. Recently, he visited poultry unit of JNKVV Jabalpur and brought about 100 chicks of Krishna-J breed. He distributed it among villagers.

He concluded that KVK intervention has reminded him impact of Green Revolution. It has created job opportunity for the landless, ensured food security and opened the door of profit making involvement for the rural youth.

Manoj Lokhande

Lokhande is young, energetic and rich farmer. He possesses 38 acres of land of which 25 acres are under irrigation. The main source of irrigation is well and three underground pipelines from dam. He also runs farm machinery workshops at field with all agricultural implements like three tractors, power thresher, power sprayer, mould bold plough, cultivator, hoe and seedd-drill. He has availed support of KVK scientists to a great extent. Subsequently, there is tremendous improvement in his economic condition. Because of improvement in economic condition he deposited Rs. One lakh in the Cooperative Bank at Jaam. Now he has all the household assets like TV, VCR, Mixer, Gezer, Refrigerator, Cooler and Fan, Dinning table, Sofa-set and Telephone connection.

Lokhande strongly criticized views of some of the villagers that KVK representatives only make visit to the farms of big farmers. Small land holders, in many cases, are incapable to avail benefits of KVK only because of poverty and resourcelessness and not because of any fault of KVK functionaries. He is aware of the market dynamics and such awareness helps him to avail maximum benefit. To him agricultural modernization has led to creation of wage opportunity at the village level but the wage rate is static and also it is regulated by demand and supply dynamics. Under the given scenario women are paid low wage than men and most often they are assigned labour intensive tasks such as weeding, transplantation and harvesting. Further, such women are neither exposed to post-harvest work nor to the related technology. Low wage and adoption of new technology led displacement of some of these workers in

Jaam village have negatively affected their income and dietary habit. They are incapable to obtain and consume new crops like carrot, garlic, coriander, brinjal and chilly. He criticized KVK for neglecting views and opinions of youth in the total planning and management of agricultural activities in Jaam village as per the requirements of these youth. However, he boldly admitted the role of KVK in the overall agricultural and social development of Jaam.

Narayan Pawar

Pawar, aged 38 years hails from Chargaon village. He has 15 acres of land of which 12 acres are well irrigated. Also, he cultivates 8 acres of land on rental (contract) basis. The process of social capital formation began early in his life. He received education up to high school level. He formed Nehru Yuva Mandal (NYM) at the village under Chhindwara based Nehru Yuva Sangathan. He was elected as Sarpanch at the age of thirty. Currently he is Janpad member. He recalled that in 1996-97 KVK scientists visited Chargaon and identified this village under Institutional Village Linkage Programme (IVLP). An intervention module, in consultation with villagers, was prepared by these scientists. Problems relating to under or low yielding of crop like gram, pea, soya etc. were documented and on this basis suitable techniques were suggested.

He said that initially NYM members had to face lots of problem and hardship because most of them were untrained. The KVK scientists inspired, helped and trained youth to adopt improved techniques for high yielding. It was because of KVK scientists that he along with other villagers produced Barseem fodder and its seed. This development made them not only sufficient in fodder production but also it improved their economic condition. It had also positive impact on health status of their cattle. Currently, he is using 10 technologies as provided/suggested by KVK. These are related to seed treatment, timely sowing, space maintenance, use of balanced fertilizer, spray of insecticides and fungicide. He is also using nadep compost, vermin compost, HYN seed and use of weedicides. Adopting of all these techniques has contributed to an increase in income and subsequently economic development. And because of economic advantage he has constructed a house and got telephone connection. His social status have improved and today many villagers follow his advice in day-to-day matters.

Pawar said that by adopting modern variety and better crop production techniques, yield of soya, gram and pea can further be enhanced. Farmers traditional practice of broadcast seeding have resulted in low and uneven crop stand and ultimately poor yield. On Farm Testing (OFT) events were conducted at different test locations to find out appropriate methods of stand establishment. Line Seeding (LS) has proved to be a better alternative for better yielding. At Chargaon soybean yield has increased from 45 to 110 per cent due to such technological initiatives. Till date he has participated in number of training programmes related to agro-forestry, dairy production and barseem production. He was facilitated by KVK in 2002 on the occasion of Kisan Samman Day.

In response to a question that when he came in contact with KVK he said that in 1999 he came in contact with KVK officials. They encouraged him to take part in Front Line Demonstration (FLD) of pea and wasteland development through agro-forestry. After attending this training-cum-demonstration programme and adopting its suggestions, he earned a lot which subsequently changed his socio-economic status. It improved his quality of life. Today he is producing pea in huge quantity and growing bamboo, eucalyptus and sisham on most part of his wasteland.

Bhadu Bhagade

Bhadu, 43 years old man of Chargaon gained economic mileage under the leadership of KVK and with the help of new technology he substantially produced gram, soya, pea and barseem. He has 8 acres of land of which 5 acres are irrigated. Every now and then he seeks help of KVK functionaries. He is little bit innovative in his ideas and well aware of the fast changes which are taking place in farm activities. Once he tried sugarcane cultivation but he could not succeed and faced monetary loss due to an attack of termite on sugarcane. He has established a mango orchard and keeps engaged in multiplication of soybean, gram and arhar (pigeon pea) and vegetables like cauliflower. He meets requirement of KVK honestly.

When he came to know from villagers that their farms are not visited by the scientists he politely requested them to make visit. He immediately before sowing organized meeting of farmers which was addressed by KVK representatives. Bhadu has leadership quality. Not only he himself has used the modern techniques but also he has suggested other farmers to use modern seeds of all varieties and to arrest spread of infestation of disease with the help of insecticides. He has availed maximum benefit but even today he is of the view that KVK personnel mainly visit the roadside farms which in most cases are owned by big farmers.

On the basis of description of some of the case studies, number of conclusions may be drawn. The analysis reveals that KVK is fruitful and functional for almost the entire village. It has not only encouraged villagers to invest in agriculture but also it has brought overall improvement in their socio-economic life. It has discouraged out-migration and encouraged production of vegetable and other cash crops exclusively for market. This has further increased frequency of their visit to local towns like Chhindwara and Nagpur. Such visits has widened their area and frequency of interaction. The overall impact of KVK exposure can easily be seen on issues like health and education, saving and expenditure, food and dress and possession of different artifacts. Change and improvement in these issues have not taken any social cost. However, it has encouraged competition among farmers although degree of co-operation is still as strong as it was during the pre-KVK intervention days.

CHAPTER – NINE

Conclusion

Broad purpose of the study was to know introduction, structure and functions of Krishi Vigyan Kendra (KVK) in select underdeveloped villages falling under the underdeveloped area and to see its impact on farm production, change in cropping pattern, income and expenditure, saving and indebtedness and finally on social institutions and cultural practices.

The study was conducted in three villages of Chhindwara district of central Madhya Pradesh, namely Simaria, Jaam and Chargaon where KVK intervention was made. All the households of these three villages participated in the study. Semi-structured interview schedule was the main instrument of data collection. However, review of secondary source information, case study and photography technique also played crucial role in data collection. Mainly male head of the households participated in the study. Some of the main findings of the study are as under:

All the three villages under study are numerically dominated by lower castes. The villages are underdeveloped in terms of availability of modern infrastructural facilities and indicators of physical development. Land is the main or the major source of livelihood for majority of villagers no matter whether they are landowners or landless. The problem of landlessness is not very serious in all the three villages. Due to the introduction of KVK technology there is an improvement in economic condition of villagers. This is apparent from the fact that during last few years about 77 per cent villagers have renovated their old houses. Further there is significant increase in both household and agricultural assets, both traditional and modern, primarily due to improvement in income due to KVK introduction. Villagers are also of the view that in comparison to the past today there is an improvement in economic condition of villagers both at the household and at the village level.

Villagers at large know about the presence and functioning of KVK in their village. They also know about broad objectives of KVK. Majority of them have availed inputs and benefits made available by the KVK. As a result, there is change in the cropping pattern in all the three villages. Production of cash crop is now common in villages. Villagers have perceived visit of KVK representatives, frequent training and field demonstration, site visit etc., as a facilitating factor for change in cropping pattern. But a significant number of them expressed that KVK should develop some specific package exclusively for the poor and the marginalized who are even today at the margin of the development process and KVK is not very much fruitful to them. To them KVK have reached to the villagers unevenly. Rich have gained mere in comparison to their poor counterpart. But it has been successful in diffusing its philosophy and agenda among majority of villagers. It has increased income and decreased indebtedness. People now take more loan for agricultural modernization from institutional sources than what they used to take for consumption purpose from non-institutional sources. There is no space for the non-institutional sources for this purpose. As a result, there is overall improvement in socio-economic condition of villagers at the household and at the village level.

Due to improvement in income and ever increasing outside contact because of frequent visit to Chhindwara, Nagpur and Bhopal cities primarily for marketing of their agricultural produce, they are now no more very much interested in seasonal out-migration. In other words, KVK have discouraged seasonal out-migration. Today they invest more in agriculture in comparison to the past. It has enhanced their income which in turn have promoted consumer culture, children's education, expenditure on health etc. in a positive manner. The type of wind which is blowing in all the three villages, due to KVK initiatives and villagers frequent visit to urban centres, have discouraged both the practice of untouchability and cases of crime. But the wind has contributed to dowry system. Today villagers give and receive more dowry. This is because of an improvement in their socio-economic status. Also it has increased age at marriage. Economic disparity among different social groups have also increased. Landless have gained only in the notional sense. Hence, from poor's view point both KVK and green revolution are similar to each other. Farm worker's income from wage work have increased but they are not paid as per the rule. Actually they are underpaid. Hence, it may be said that KVK have contributed to an improvement in income of many but from social change point of view its impact is mixed. It has contributed to an institutionalization of democracy as well as strengthening of tradition.

Policy Implications

In order to ensure equitable, cost effective and sustainable functioning of KVK particularly from stability, productivity and equity point of view, number of suggestions were expressed by cross section of villagers. If we integrate and critically examine all these suggestions, following points may be drawn for policy implications:

1. Some of the villagers said that they live in extreme corner of the village where KVK representatives do not make visit to their fields. As a result, they do not know about importance of KVK which in turn inhibit them to use its benefits, which might have been otherwise availed by them. Hence, it is suggested that KVK functionaries should visit all the farm lands, no matter where they are located, who are their owners and what are the types and quality of land and soil. It must reach to each and every household in the experimental villages. By taking this step definitely the number of beneficiaries will increase and people will form positive opinion about KVK and its functioning. Hence, the project must be inclusive and equitable.
2. It is observed that many of the landless households and landless wage earners are not getting proper and direct,

benefit from KVK. The difference between pre-KVK days and today is that now they get more man-days employment because of change in the cropping pattern and farmers increasing inclination in favour of farm work. If this criticism is valid then it may be said that KVK is not different than the Green Revolution particularly from poor's point of view. KVK should introduce some specific programmes exclusively meant for such poor and downtrodden villagers, most of whom are landless and they are unable to avail benefits of KVK directly. They may be involved in some value-added manufacturing activities. Also, some training programmes may be organized exclusively for these poor and landless population. So long as poverty eradication measures are not taken they cannot come under the direct and effective umbrella of KVK.

3. Currently there is no frequent interaction between KVK representatives and the Gram Panchayat. It appears that both these institutions are isolated compartments of the train which is going to the same direction and destination. Looking to the increasing power, role and importance of gram panchayat as well as its elected representatives, the amount of interaction between all the change agents operating at the village level should be increased and it must be institutionalized. KVK can take help from them in the selection of both field and the farmers, etc. for experiments. Suggestions relating to the effective interface between KVK and panchayat needs to be properly documented.
4. From earlier studies it is obvious that with an increase in the number of development projects at the village level during last several decades not only the bases of factions have shifted from caste, lineage and area of habitation to development issues but also number of factions have increased. Therefore, while making experiments KVK must know these factions, take faction leaders into confidence and organize meetings with them on common places like schools, panchayat buildings etc. on frequent basis. This effort will encourage larger participation otherwise like

other development projects KVK will also suffer not on the basis of its merits and demerits but due to faction ridden politics.

5. In course of the study it was experienced that there is decrease in the production of some of the traditional crops during last couple of years. This is a general trend and the study area is not an exception. If this trend continues it can affect credibility of KVK negatively. KVK is not directed to promote irrigation and fertilizer centric cash crop at the cost of environment friendly and mass oriented traditional crop which is produced mainly for consumption purpose. Hence, KVK should investigate reasons and suggest feasible ways and means to overcome this problem and encourage production of cash crops as well as traditional crops. It is needless to mention that traditional crops have historically adjusted with local conditions and ecology. This factor should not be overlooked while subscribing to new crops. In other words, while subscribing production of new crops, proscriptions and prescriptions should be prepared and made available to farmers. Carrying capacity of different categories of farmers should also be taken into consideration.
6. Since villagers are from different social and economic backgrounds, some are landless and some are the big farmers, some are under debt and some are money-lenders, therefore, for logistic reason they all cannot come under one umbrella to attend meeting or training programme. Infrastructure constraint can also prohibit some of them to sit and attend programme at one place. Hence, it is suggested that a small group of farmers, small group of agricultural workers, group of women, etc., should be formed and they should be trained separately and jointly at the local level itself keeping in view the social and political undercurrents in mind. The Farmers Club of both men and women should be formed and every now and then meetings and demonstration programmes, etc., should be organized. The KVK agents as well as other change agents from village itself should be equipped with modern agricultural practices and

they should be motivated to diffuse it among villagers. If some of the villagers are assigned this work the degree of success of KVK initiatives can be enhanced because change agents are informally known to every villager.

7. While organizing training programme, some specific precautions should be taken. Looking to the memory power of villagers such programmes should not only be organized at the village level itself, it should be practical in nature but also it should be organized immediately before the implementation of such programmes. In other words, in the case of harvesting they should be trained how to harvest effectively and manage their land immediately after harvesting. In the case of water harvesting they should be trained during the rainy season itself about cost-effective techniques of water harvesting and its use. Time of training, duration of training, frequency of training and content of training etc., should be decided in consultation with the villagers so that maximum number of them can feel at ease and can attend the programme and can really get benefit out of it. Also, while introducing relatively complex and costly technology precautions relating to its adoption, maintenance etc., must be taken otherwise once non-functional it will remain non-functional for want of its repair at the local level. There are examples of hand pump and tube-wells from the study area which remained out of order for months for want of suitable repair.
8. The KVK should prepare exhaustive case study of some of the successful farmers highlighting factors which enabled them to succeed and distribute it among villagers not only in the villages where training and demonstration was conducted rather in the neighbouring villages also so that people at large can learn lesson from success stories. In this case help of radio, newspaper, mobile phone and electronic media can also be taken. Even the successful farmers may also be invited to interact with villagers and the farmers attending training programmes.
9. Quality of seed, quality of fertilizer, quality of soil, etc. are

directly linked with production. For this purpose farmers need support from KVK. Those who have traditional seed they need modern seed and its protection and preservation technique. Those who do not have such seed they need HYV seed but many times they do not have access to seed in time. For example, if they want to grow papaya they do not know from where to obtain quality ridden seed. Hence, it is suggested that KVK should establish Farmers Cooperative Seed Bank (FCSB), if not at the village level then at least at the Block level. In certain cases seed should be provided on credit. Some progressive farmers should be assisted by the KVK to produce quality seed at the local level for its local use. Establishment of Seed Bank (SB) or Seed Village (SV) was also an idea suggested by some of the villagers. This problem may also be mitigated by forming Farmers Cooperative Society (FCS). Responsibilities of similar nature may also be entrusted upon Self-Help Groups (SHGs) now found in thousands in villages.

10. To ensure high productivity and maximum use of new technology proper and suitable investment in land is also required. Farmers can invest in land in different ways during different seasons as per the requirement of specific crop as suggested by KVK representatives. This intervention requires monetary support to farmers. Hence, besides scientific and technical support, financial support should also be made by the government. If KVK link itself with already undergoing government sponsored rural development schemes which are today in significant number, the output may be praiseworthy. They can be associated with watershed project and other projects of similar nature.
11. Improvement in farm production is not a sufficient condition for economic well-being of farmers and villagers. Skilful marketing of farm produce is also an important condition. Individual farmers cannot work as per the requirements of the existing and the emerging market forces. Many times many of them even do not know the

nature and functioning of market forces. As a result, they cannot sell their produce in a profitable manner and therefore, maximum part of the benefits are manipulated by traders and middle men. They are almost left with empty hands. To ensure better marketing and exploit market opportunity and discourage middlemen, co-operative societies of farmers in general or co-operative society of garlic producing farmers, etc., in particular, as an example, should be formed and government should come forward to assist them. Success of white revolution in different parts of the country and its management and promotion by co-operative societies is an example of success story which may also be replicated in the case of other farm produce.

12. It is observed that Chhindwara unit of KVK is understaffed. More or less similar is the situation with other *Kendras* today. And as a result, not only they are overburdened but also they may have to compromise with some essential activities. This contributes to unnecessary delay in the execution of development programmes. Hence, it is suggested that sufficient staff should be made available to the centre at the earliest. But if it is not feasible due to the shortage of fund, then village youth, panchayat representatives and SHG members should be trained to undertake this responsibility and they should be provided reasonable honorarium. And for this purpose in mind these change agents should be imparted orientation and training on regular basis.
13. If value added manufacturing centres are established at the local level and cash crops are produced in a scientific manner with proper use of traditional knowledge it will create employment opportunity which may further discourage seasonal out-migration which in turn will fasten the ongoing process of development. If out-migration is discouraged by providing suitable employment opportunity at the village level, people at large will participate in the ongoing rural development projects. Also if out-migration

is discontinued it will create pressure on the local level change agents to deliver properly.

14. Currently, there is no facility available for testing of soil at the KVK level. This facility should be made available free of cost at the Block level. Farmers must have a list of paper of all the pieces of land owned by them and features of all the individual plots from specific crop point of view. For this purpose a diary containing fundamental information about each and every aspect of farming should be made available to individual farmers, free of cost, with a number of prescriptions and proscriptions.

In short, if these suggestions are taken into consideration by all the concerned it will create higher amount of interest in farming, discourage out-migration, contribute to high production, proper marketing of farm produce and finally an improvement in economic condition. It will reduce social and economic disparity between the classes and will also create inclusive social and economic environment which will suit to all the social groups including all the components of local ecology, environment, and social structure.

Bibliography

Acharya, S.S. (1973): "Impact of technological change on farm employment and income distribution in Agriculture," *Ind. Des. Abst*: 2(3).

Adalman, Irma and Morris, Cynthia Toft (1967): *Society, Politics and Economic Development,* Baltimore: The John Hopkins Press.

Anonymous (1997): *FAO Production Year Book*, Rome.

Arora, D.R., Agarwal, B.K. and Gupta, A.K. (1989): "Green Revolution and social change in Punjab," in (ed.) *Green Revolution and Social Change*, New Delhi: Ajanta Publication, pp. 79-90.

Aslam, Mohd. (1981): *Social Implications of Technological Change in Rural Kashmir;* New Delhi: Inter-India Publication.

Bailey, F.G. (1959): *Tribe, Caste and Nation,* Bombay: Oxford University Press.

Bhattacharya, S.K. (1976): *Farmers Rituals and Modernisation;* Calcutta: Minerva Associates Publications.

Bhattacharya, Dhiros (1970): *The Indian Plans,* Calcutta: Progressive Publisher, p. 19.

Chand, Ramesh; Sidhu, D.S. and Kaul, J.L. (1985): "Impact of Agricultural Modernization on labour use pattern in Punjab with Special reference to women labour;" *Indian Journal of Agricultural Economics;* 40 (3); 252-254.

Chaudhary, A.P. (2002): "Economic Reforms and Agricultural Development in Rajasthan;" *Indian Journal of Social Research;* 43 (2): 113-125.

Chaudhary, S.N. (1990): *Bio Gas Technology in Development of Rural Women;* New Delhi: Amer Prakashan.

Chaudhary, S.N. (1999): *Power-Dependence Relations: Struggle for Hegemony in Rural Bihar*, New Delhi: Har-Anand.

Chaudhary, S.N. (2003): *Social and Economic Impact of Chernoi Land on Dalits: Case Study of Raisen District,* (mimeo) Bhopal: Tribal Research Institute.

Chaudhary, S.N. (2012): *In Search of Sustainable Development: Scenario in a Tribal Village of Madhya Pradesh*, New Delhi: Serials Publications.

Coale, Ansely J. and Hoover Edger (1958): *Population Growth and Economic Development in Low Income Countries;* New Jerry: Princeton University Press.

Dantwalla, M.S. (1969): "Agriculture in a Developing Economy—the Indian Experience;" *Rural Sociology in India;* (ed.) A.R. Desai, Bombay: Popular Books.

Dasgupta, S. and Madan, G.R. (1978): *Community and Agriculture in two Indian Villages;* Calcutta: Editions Indian.

Desai, A.R. (1969): *Rural Sociology in India;* (fifth edition), (ed.), Bombay: Popular Books.

Dev, P.C. and Singh R.C. (1975): *Technology and Social Change in Rural Punjab;* Technical Report, Department of Economics and Sociology: Punjab Agriculture University; Ludhiana: 75.

Dube, S.C. (1974): *Contemporary India and its Modernisation;* New Delhi: Vikas Publishing House Pvt. Ltd.

Dubey, D.S. (1925): quoted by Gungnnlla N.N. in his paper read before the East India Association in 1925 edited in *Problems of Rural India* by N.N. Gungnnlla.

Frankel, Francine R. (1971): *India's Green Revolution: Economic Gains and Political Costs;* Bombay: Oxford University Press (Indian Edition): 193.

Gangwar, A.C. (1989): "Impact of Green Revolution on Consumption pattern and Nutritional Status" *in Green Revolution and Social Change;* (Ed) Sharma, M.L. and Dak, A.M.; New Delhi; Ajanta Publication.

Gill, D.K. and Saini, S.K. (1987): "Social Implications of Green Revolution"; paper presented at ICSSR, Seminar on Social, Economic and Political Implications of Green Revolution in India; P.A.U., Ludhiana.

Haque, M.A. and Khan, M.H. (1999): "Impact of Barind Multiple Development Project on Socio-economic development in Rajshahi region;" *Bangladesh Journal of Training and Development;* 12: (122) 25-31.

Jain, Uday and Mishra, Girishwar (1996): "Image of Science and Technology in Village and Process of Social Change" in (Ed): *Law, Justice and Social Change;* D.R. Saxena; New Delhi: Deep & Deep Publications: pp. 313-319.

Khan, Alimur Haque (1997): "Adoption of new agricultural technology, and economic Change in Tharus of Uttar Pradesh;" *Social Change:* March-June; 27 (1&2): 243-253.

Khan, Khader H. and Chandregowda, K.N. (2000): "Impact of Whole Village Development Project at Harati Village of Kolar Taluk in Karnataka;" *Financing Agriculture,* New Delhi.

Khan, M. and Khan, I.M. (1999): "Role of KVK dissemination of new technology among the farmers:" *Rural India,* June: 163-166.

Kumar, B. and Singh D.P. (1994): "Profile of KVK Trainers;" *Indian Journal of Extention Education,* 30 (1&4): 42-49.

Kuppuswamy, B. (1990): *Social Change in India;* New Delhi: Konark Publisher Pvt. Ltd.

Mahalingam, N. (2002): "Strategic Approach for Poverty": *Kisan World*: Janury; 2002: 5.

Maheshwari, B.P. (1971): *A Critical Appraisal of Economic Development Between (1914-1947): Industrial and Agricultural Development of India Since 1914*: New Delhi: S. Chand & Co.

Malasis, L. (1975): *Agriculture and Development Process: Tentative Guidelines for Teaching*, Paris: UNCESCO.

Mangat, Remandeep and Gill, L.S. (1999): "Krishi Vigyan Kendra: Creating Self-Employment and Generating Income in Punjab;" *Kurukshetra*: 47 (4): 35-38.

Meena, R.K. and Chauhan, M.S. (2002): "Constraints as perceived by the farmers in Adoption of improved production of Groundnut," *Indian Journal of Social Research*: 43(3): 201-207.

Mehta, B.R. (1964): Inaugural address to Indian Society of Agriculture Economic in 1964.

Minhas, B.S. (1969): *Forth Five-Year Plan: Objective and Policy Frame*; New Delhi: Commerce pamphlet: 21-24.

Mishra, B.N. and Pandey, Vandana (1992): "Role of Agriculture in the rural Development; A case study of Mirzapur District of U.P.;" *Geographical Review of India*; 45 (1): 37-49.

Mohanty, B.B. (1996): "Agriculture Modernization in Rural Orissa: Prosperity and Destitution;" *Man In India*; March; 76 (10) 81-90.

Mohanty, B.B (2000): "Agriculture Modernization in Rural Orissa: Land Transfer and Ownership Pattern:" *Sociological Bulletin*; March 2000: 49 (I): 63-90.

Myrdal, Gunnar (1968): *Asian Drama: An Enquiry into the Poverty of Nations*, 2, London: The Penguin Books.

Nair, P. K. K. (1990): "Reaching out to the needy;" *The Hindu Survey of Indian Agriculture*: 29-32.

Narayanamoorthy, A. (1992): "Impact of tractor and weedicides on yield and Employment in Sugarcane Cultivation". *Journal of Rural Development*; NIRD, Hyderabad; 2: 351-362.

Nichols, William (1964): *The Place of Agriculture in Economic Development* (ed.) by Eicher and Witt, New York, Mc Graw-Hill Book Company: p. 19.

Panda, R.K.(1996): "Agriculture Tenancy and Resource use efficiency: A micro level Analysis": *Journal of Rural Development*; 15 (2): 187-200.

Paroda, R.S. (1999): "For a food security future;" *The Hindu Survey of Indian Agriculture*; 17-21.

Prased, C. (1990): "Krishi Vigyan Kendra. A Land Mark for Vocational Training and First Line Technology Transfer;" *Mah. Joun. Ext. Ed.*; (9) 229-235.

Rai, B.D.; Tiwari, K.B.; Thakur P. and Saxena, A.K. (1995): *Impact Study of the Transfer of Technology Programme of Krishi Vigyan Kendra, Chhindwara in adopted Village Badnoor:* KVK; Chhindwara, Publication.

Rajput, A.M. and Varma, A.R. (1998): "Flow of Institutional credit in agriculture in Indore district of Madhya Pradesh;" *Indian Journal of Agriculture Economics*; 53 (3); 512.

Raju, V.T. (1982): *Impact of New Technology on From Income Distribution and Employment*; New Delhi: National Publishing House, 107.

Randhawa, M.S. (1974): *Green Revolution*; New Delhi: Vikas Publishing House Pvt. Ltd.

Singh, Gyanendra (2002): "Equipment to ensure timeliness," *The Hindu Survey of Indian Agriculture;* 195-201.

Singh, Y. (1976): "Historicity of Modernization," in (ed.) *Tradition and Modernization: Processes of Continuity and Change,* S.K. Srivastava Allahabad: Indian International Publications.

Sabharwal, Deshraj (2001): "New Agro Technology and Change": *Man in India*: 81 (122): 179-195.

Sachchidananda, (1972): *Social Dimensions of Agriculture Development*; Delhi: National Publishing House.

Sangram Singh, S.P.; Kanungo, A.P. and Mahapatra, B.P. (1993): "Impact of KVK on socio-economic and infrastructure utilization of farmers"; *Orissa Journal of Agriculture Research;* 6 (1&2): 31-38.

Sen, B.R. (1970): "Opportunities in the Green Revolution": *Economic and Political Weekly*; March 28: A 8-21.

Sharma, H.R. and Diwakar, Ashok (1989): Transfer of Technology and Response of Small Farmers: *Margin*: January-March: 21(1):45-42.

Sharma, Ramawtar and Sharma, S.K. (2000): "Role of KVK in Dissemination of latest Agriculture Technology Among the Farmer Communities"; *Indian Journal of Social Research*: 42 (1&2): 69-73.

Sharma, T.N.; Sharma, D.V. and Sharma, V.B. (2000): "Knowledge and adoption levels of Beneficiaries and Non-beneficiaries of KVK—Crop production technology"; *Madhya Pradesh Journal of Extension Education;* II & III (2); 45-46.

Sharma, T.N.; Sharma, V.B. and Sharma, D.K. (1996): "Factors associated with knowledge and adoption of vegetable production technology by beneficiaries and non-beneficiaries of KVK, Chhindwara"; *JNKVV Research Journal*; 30 (1&3): 44-46.

Shergill, H.S. and Singh, Gurmail (1995): "Poverty in Rural Punjab: Trend over Green Revolution Decades": *Economic and Political Weekly;* (June 1995) A -80-83.

Singh, Srinath (1976): *Modernization of Agriculture: A Case Study in Eastern Uttar Pradesh;* New Delhi: Heritage Publisher.

Srinivas, M.N. (1982): *Indian Social Structure:* Delhi: Hindustan Publishing Corporation.

Supe, S.V. (1987): *An Introduction to Extention Education:* New Delhi: Oxford and IBH Publishing Co. Pvt. Ltd.

Swaminathan, M.S. (1989): "Agricultural Planning Improving the Quality of Life"; *Indian Farmer Timer;* (6), 11.

Swaminathan, M.S. (1999): "Dr. Swaminathan Committee Report on Education for Agriculture": *Agriculture Education in India,* ICAR, New Delhi, 3-7.

Taneja, N.K. (1994): "Technological Change in Agriculture and Output-employment Linkages: An Econometrics Analysis"; *Asian Journal of Economic and Social Studies;* 13 (4) (Supplementary); 315-359.

Thakur, C.L.; Sharma, K.G. and others, (2000): Adoption of advanced Agriculture Technologies by Marginal and Small Farm Families of Mandsour District in Madhya Pradesh": *Indian Journal of Agriculture Economic;* 55: (3), 532.

Thakur, P.; Sharma, R.C. and Tiwari, K.B. (1998): "Front Line Demonstration introduction to technology of Chickpea Production"; *Agriculture Science Digest*: 18 (4) 223-224.

Tharan, Thi. UT.; Mahabub Hassain and Janaiah, A. (2000): "Modern farm technology and infrastructure in Vietnam: Impact on Income distribution and Poverty": *Economic and Political Weekly;* December 35 (52 & 53), 4638-43.

Thekkamalai, S.S. (1983): *Social Change and Modernization: Rural Development and Social Change in India:* Delhi: D.K. Publication.

Yadav, V.P.; Narwhal, R.S. and Yadav, H.D. (2000): "Organizing Vocational Training Programme in Krishi Vigyan Kendra," *Haryana Farming;* 18-21.

Index